TEXTS AND READINGS IN MATHEMATICS **67**

Connected at Infinity II

Connected at Infinity II
A Selection of Mathematics by Indians

Edited by

Rajendra Bhatia
C. S. Rajan
Ajit Iqbal Singh

HINDUSTAN
BOOK AGENCY

Published by

Hindustan Book Agency (India)
P 19 Green Park Extension
New Delhi 110 016
India

email: info@hindbook.com
www.hindbook.com

ISBN 978-93-80250-51-9

Contents

Preface

In 2003 we published a collection of articles *Connected at Infinity*. The editors' aims, and their travails, were explained in the two paragraphs that we reproduce from the Preface to that volume:

The Editors invited a few persons to write an article that should explain to the nonspecialist an important piece of work done by an Indian mathematician in the mid-twentieth century. The contributions of these mathematicians have been major landmarks in the subject and have inspired and influenced a lot of later work. As such they are very well-known to all those who study and do research in certain areas of mathematics. The formidable background needed to understand their work, however, prevents many others from appreciating its significance. The articles collected here are written to help create such an appreciation.

Not all those who were invited to write for this volume agreed to do so, not all who agreed kept their promise, and not all articles we finally received are included here. Therefore, this collection is far less complete than what we had originally planned. We hope there will be a sequel that would redress this.

Ten years later the travails were forgotten, and we began work on the promised sequel. The present collection is the outcome. This too is far less complete than we planned, and for the same reasons. We do hope that it will be of as much interest as the first collection, and that this too will have a sequel.

We thank our authors for the huge amount of time and effort they spent in writing their pieces, and for patiently waiting while we got the collection together. We thank several colleagues who read the drafts of these articles and made suggestions for changes.

The first collection was compiled to mark 10 years and 25 volumes of the TRIM series. The series has since then grown with the support of several authors, reviewers, editors and readers. We record our thanks to them, and to the National Board for Higher Mathematics for their continued support.

Orthogonal Latin Squares and the Falsity of Euler's Conjecture

Aloke Dey*

1. Introduction

The problems relating to the existence and construction of orthogonal Latin squares have fascinated researchers for several centuries now. Though many important discoveries have been made, some problems still remain unresolved. Latin squares and orthogonal Latin squares have a beautiful underlying structure and are related to other combinatorial objects. These have applications in different areas, including statistical design of experiments and cryptology. Comprehensive accounts of the theory and applications of Latin squares are available in the books by J. Dénes and A. D. Keedwell (1974, 1991) and C. F. Laywine and G. L. Mullen (1998).

One of the most intriguing questions about orthogonal Latin squares was raised in the 18th century. In 1782, Leonhard Euler made a famous conjecture that there does not exist a pair of orthogonal Latin squares of order $4t + 2$, where $t \geq 1$ is an integer. The first serious doubts on the truth of Euler's conjecture were raised by R. C. Bose and S. S. Shrikhande on the one hand and by E. T. Parker on the other, both appearing in 1959. Subsequently, all of them collaborated to prove that Euler's conjecture was indeed false for every integer $t > 1$. This is considered as a landmark result in combinatorial designs. In this article, we present a brief survey of some of the important results on orthogonal Latin squares with emphasis on the work of Bose and Shrikhande leading to the falsity of Euler's conjecture. In the last section, we mention some historical facts related to orthogonal Latin squares.

2. Orthogonal Latin Squares

A Latin square of *order* (or, side) s is an $s \times s$ array with entries from a set of s distinct symbols (or, letters) such that each symbol appears in each row and

---

*Indian Statistical Institute, New Delhi

each column precisely once. Two Latin squares of the same order are said to be *orthogonal* to each other if, when any of the squares is superimposed on the other, every ordered pair of symbols appears exactly once. Orthogonal Latin squares can be equivalently defined as follows: Two Latin squares of side s, $L_1 = (a_{ij})$ on symbols from a set S_1, and $L_2 = (b_{ij})$ on symbols from a set S_2, are orthogonal if every element in $S_1 \times S_2$ occurs exactly once among the s^2 pairs (a_{ij}, b_{ij}), $1 \le i, j \le s$. For example, consider the following pair of Latin squares of order $s = 4$:

$$L_1 = \begin{array}{cccc} A & C & D & B \\ B & D & C & A \\ C & A & B & D \\ D & B & A & C \end{array}, \quad L_2 = \begin{array}{cccc} \alpha & \delta & \beta & \gamma \\ \beta & \gamma & \alpha & \delta \\ \gamma & \beta & \delta & \alpha \\ \delta & \alpha & \gamma & \beta \end{array}.$$

Superimposing L_2 over L_1, one gets the following arrangement:

$$L = \begin{array}{cccc} A\alpha & C\delta & D\beta & B\gamma \\ B\beta & D\gamma & C\alpha & A\delta \\ C\gamma & A\beta & B\delta & D\alpha \\ D\delta & B\alpha & A\gamma & C\beta \end{array}.$$

Clearly, L_1 and L_2 are orthogonal to each other. The arrangement like L is called an *Eulerian square*, named after the legendary mathematician Euler (1707–1783). Euler studied such objects in 1782 and also made a famous conjecture about their existence.

If in a set of Latin squares every pair is orthogonal, then the set is said to form a set of *mutually orthogonal Latin squares* (MOLS). It is a nice exercise to show that the number of MOLS of order s is bounded above by $s - 1$. When this upper bound is attained, we say that there is a *complete set* of MOLS.

Orthogonal Latin squares are related to other combinatorial objects and we describe some of these. Latin squares are related to quasigroups. Recall that a *quasigroup* is a pair $< Q, * >$, where Q is a set and $*$ is a binary operation on Q, such that the equations $a*x = b$ and $y*a = b$ are uniquely solvable for every pair of elements $a, b \in Q$. It is easy to see that the Cayley (or 'multiplication') table of a quasigroup (with the headline and sideline removed) is a Latin square. Two quasigroups $< Q, \odot >$ and $< Q, \oplus >$ with binary operations $\odot$ and $\oplus$, defined over the same set Q are said to be *orthogonal* if the equations $x \odot y = z \odot t$ and $x \oplus y = z \oplus t$ together imply that $x = z$ and $y = t$. When a pair of quasigroups $< Q, \odot >$ and $< Q, \oplus >$ are orthogonal, their corresponding Latin squares are also orthogonal. A pair of orthogonal Latin squares, derived from

2

two orthogonal quasigroups with 3 elements is shown below:

$\odot$	1	2	3
1	1	3	2
2	2	1	3
3	3	2	1

$\oplus$	1	2	3
1	1	2	3
2	2	3	1
3	3	1	2

Another object related to orthogonal Latin squares is an *orthogonal array*. An orthogonal array $OA(N, k, s, g)$ of strength g ($2 \le g \le k$) is a $k \times N$ matrix A with entries from a finite set S containing $s \ge 2$ elements, such that in any $g \times N$ sub-matrix of A, each of the g-tuples of symbols from S occurs the same number of times, say λ times, as a column. It follows then that in an $OA(N, k, s, g)$, $N = \lambda s^g$. The integer λ is called the *index* of the array.

Let $\{L_u, 1 \le u \le k\}$ be a set of k MOLS of order s on symbols $1, 2, \ldots, s$. Form a $(k+2) \times s^2$ array $A = (a_{ij})$ whose columns are

$$(i, j, L_1(i, j), L_2(i, j), \ldots, L_k(i, j))' \text{ for } 1 \le i, j \le s.$$

Then, one can show that A is an orthogonal array $OA(s^2, k+2, s, 2)$ of strength *two* and index unity. Conversely, reversing the above steps one can obtain k MOLS of order s from an $OA(s^2, k + 2, s, 2)$. As an example, consider the following three MOLS of order 4:

$$L_1 = \begin{matrix} 1 & 2 & 3 & 4 \\ 4 & 3 & 2 & 1 \\ 2 & 1 & 4 & 3 \\ 3 & 4 & 1 & 2 \end{matrix}, \quad L_2 = \begin{matrix} 1 & 2 & 3 & 4 \\ 3 & 4 & 1 & 2 \\ 4 & 3 & 2 & 1 \\ 2 & 1 & 4 & 3 \end{matrix}, \quad L_3 = \begin{matrix} 1 & 2 & 3 & 4 \\ 2 & 1 & 4 & 3 \\ 3 & 4 & 1 & 2 \\ 4 & 3 & 2 & 1 \end{matrix}.$$

Following the steps of construction just described, one gets an $OA(16, 5, 4, 2)$, displayed below.

$$\begin{bmatrix} 1111 & 2222 & 3333 & 4444 \\ 1234 & 1234 & 1234 & 1234 \\ 1234 & 4321 & 2143 & 3412 \\ 1234 & 3412 & 4321 & 2143 \\ 1234 & 2143 & 3412 & 4321 \end{bmatrix}.$$

As stated earlier, the maximum number of MOLS of order s is $s - 1$. This upper bound is attainable if s is a prime or a prime power. The problem of construction of a complete set of MOLS of order s when $s = p^n$ where p is a prime and n is a positive integer was solved by Bose (1938) and independently, by W. L. Stevens (1939). We briefly describe below the method given by Bose (1938).

Let s be a prime or a prime power and let $GF(s)$ denote the Galois field of order s with elements $\rho_0 = 0, \rho_1 = 1, \rho_2, \ldots, \rho_{s-1}$. Consider a square L_i, whose

(r, t)th cell is filled by the element

$$\rho_r + \rho_i \rho_t, \ \ 1 \le i \le s - 1, \ 0 \le r, t \le s - 1.$$

It can be verified that (i) L_i is a Latin square of order s for each i and (ii) for $i \ne j$, L_i and L_j are orthogonal. This then provides a method of constructing a complete set of MOLS of order s, where s is a prime or a prime power.

3. Connection with Finite Projective Planes

Bose (1938) showed that there is a 1-1 relation between a complete set of MOLS and a finite projective plane. Consider a (finite) set of elements, called *points*, certain subsets of points, called *lines* and an *incidence relation* between them. The set of points and lines are said to form a *finite projective plane* if the following axioms hold:

P1: There is exactly one line which is incident with every pair of distinct points, i.e., there is exactly one line through a pair of distinct points.

P2: There is exactly one point in common with every pair of distinct lines.

P3: There exist four points, no three of which are incident with the same line.

Note that there is a symmetry in the axioms P1–P3, making the principle of duality hold in the geometry in the sense that the roles of points and lines can be interchanged, without affecting the properties of the geometry. In a finite projective plane, each line is incident with $(s + 1)$ points and each point is incident with $(s+1)$ lines, where $s \ge 2$ is an integer. The total number of points, as also the total number of lines, in a finite projective plane is (s^2+s+1). A finite projective plane in which each line has $(s + 1)$ points is said to be of *order* s.

A finite projective plane of order $s(\ge 2)$ where s is a prime or a prime power can be constructed as follows. Let $s = p^q$, where p is a prime and q, a positive integer. An ordered triplet (x, y, z), where $x, y, z \in GF(s)$ and $(x, y, z) \ne (0, 0, 0)$, is said to define a point of the finite projective plane. Two ordered triplets (x_1, x_2, x_3) and (y_1, y_2, y_3) define the same point if and only if $y_i = cx_i$, $(i = 1, 2, 3)$, where $(0 \ne c) \in GF(s)$. A linear homogeneous equation $ax + by + cz = 0 \pmod{s}$, where $a, b, c \in GF(s)$ and $(a, b, c) \ne (0, 0, 0)$, defines a line. Two such equations define the same line if their corresponding coefficients are proportional. The point (x_0, y_0, z_0) is said to be incident with the line $ax + by + cz$ if $ax_0 + by_0 + cz_0 = 0$. The set of points, lines and the incidence relation so defined can be verified to satisfy axioms P1–P3 and thus form a finite projective plane of order s.

We now describe the connection between a complete set of MOLS and a finite projective plane, both of order s, where s is a prime or a prime power.

4

Let ℓ be a line in a finite projective plane and $x_r, x_c, x_1, \ldots, x_{s-1}$ be the points on ℓ. The remaining $s(s+1)$ lines (other than ℓ) can be partitioned into $(s+1)$ sets of s lines each, such that any pair of lines in the same set intersect ℓ at some point. We may call ℓ and the points on it as the line and points at infinity, respectively. Let us delete the line ℓ and all points on it and, call the remaining lines and points as finite lines and finite points. Then the set of lines intersecting at say, x defines a *pencil* of finite lines, which may be denoted as $[x]$. The finite lines thus get partitioned into pencils of lines $[x_r], [x_c], [x_1], [x_2], \ldots, [x_{s-1}]$. Let the s lines in a pencil be labeled by the integers $0, 1, \ldots, s-1$. The lines of $[x_r]$ and $[x_c]$ can be associated with the rows and columns, respectively, of an $s \times s$ array, such that the (u, v)th cell of the array is the point of intersection of line u of $[x_r]$ and line v of $[x_c]$. For $1 \leq i \leq s-1$, corresponding to the pencil $[x_i]$, let us form a square L_i whose (u, v)th entry is the number corresponding to the line of $[x_i]$ which passes through the point corresponding to the cell (u, v). Then, it can be shown that L_i is a Latin square of order s. Through the point corresponding to the (u, v)th cell, there is exactly one line of $[x_i]$ and one line of $[x_j]$, $i, j = 1, 2, \ldots, s-1, i \neq j$. This shows that for $i \neq j$, the Latin squares L_i and L_j are orthogonal.

Conversely, let $L_1, L_2, \ldots, L_{s-1}$ be a complete set of MOLS of order s whose symbols, without loss of generality, can be taken to be $0, 1, \ldots, s-1$. Let L_R (respectively, L_C) be two $s \times s$ arrays in which the symbol i appears in all the cells of the row (respectively, column) numbered $i, 0 \leq i \leq s-1$. Define s^2 points $(i, j), 0 \leq i, j \leq s-1$. Join the points with the same first coordinate by a line. Then we have s lines, which are parallel (i.e., have no common point). Let these lines intersect at a new point x_R. Similarly, obtain s lines by joining the points with the same second coordinate and let these lines intersect at a new point, say x_C. For the Latin square L_i, let the s^2 cells be identified with the s^2 points described above. Join the points corresponding to the same integer in L_i and let these s lines intersect at a new point $x_i, 1 \leq i \leq s-1$. Finally, join $x_R, x_C, x_1, \ldots, x_{s-1}$ by a line. Then it can be verified that this collection of points and lines forms a finite projective plane.

As noted above, a finite projective plane of order s exists if s is a prime or a prime power and this plane coexists with a complete set of MOLS of order s. A natural question then is: for what other values of s does a finite projective plane exist? This is one of the most difficult questions in finite geometry. Unfortunately, very little is known. The earliest result about the existence of finite projective planes (of non-prime power orders) is due to R. H. Bruck and H. J. Ryser (1949). This result forms a special case of the results in S. Chowla and Ryser (1950) and Shrikhande (1950), given in different forms. We state below the result, called the *Bruck-Ryser-Chowla Theorem* in the language of finite projective planes, though an equivalent statement in terms of the existence conditions of certain balanced incomplete block designs is also used in the literature.

Theorem 3.1. *If a projective plane of order $s \equiv 1$ or 2 (mod 4) exists, then s is the sum of squares of two integers.*

Equivalently, the above result states that if $s \equiv 1$ or 2 (mod 4), then no finite projective plane of order s exists unless s is the sum of squares of two integers. Thus, Theorem 3.1 rules out the existence of a finite projective plane of orders $s = 6$ and 14. However, it does not rule out the existence of a projective plane of order 10. C. W. H. Lam, L. H.Thiel and S. Swiercz (1989) showed the non-existence of a finite projective plane of order 10 through a massive computer search. Nothing more than the above stated results on the existence of a finite projective plane of order s, where s is neither a prime nor a prime power seems to be known.

4. The MacNeish-Mann Conjecture

For an integer s, let $N(s)$ denote the maximum number of MOLS of order s. Then, as seen earlier, $N(s) = s - 1$, if s is a prime or a prime power. A challenging problem is to determine the value of $N(s)$ (or, bounds on $N(s)$) when s is neither a prime nor a prime power. One of the earliest results in this direction is due to H. F. MacNeish (1922); this was generalized somewhat and put on an algebraic foundation by H. B. Mann (1942). Let $s = p_1^{n_1} p_2^{n_2} \ldots p_m^{n_m}$ be the prime-power decomposition of s, where $p_1, \ldots, p_m$ are distinct primes and $n_1, \ldots, n_m$ are positive integers. Define

$$n(s) = min\{p_1^{n_1}, p_2^{n_2} \ldots, p_m^{n_m}\} - 1. \tag{4.1}$$

MacNeish (1922) showed that $N(s) \geq n(s)$. MacNeish went further to conjecture that $n(s)$ is also the upper bound on $N(s)$ and therefore, $N(s) = n(s)$. The MacNeish-Mann construction of $n(s)$ MOLS of order s can be described as follows: Consider the system of s elements

$$\gamma = (g_1, g_2, \ldots, g_m),$$

where for $1 \leq i \leq m$, $g_i \in GF(p_i^{n_i})$. The addition and multiplication of these elements are defined by the following rules:

$$\gamma_1 + \gamma_2 = (g_1, \ldots, g_m) + (h_1, \ldots, h_m)$$
$$= (g_1 + h_1, \ldots, g_m + h_m);$$
$$\gamma_1 \gamma_2 = (g_1 h_1, \ldots, g_m h_m),$$

where the operations above in each component are as defined in the corresponding Galois field. The system so constructed is however, not a field, as not all elements in the system have a multiplicative inverse.

Let $g_i^{(0)} = 0, g_i^{(1)} = 1, g_i^{(2)}, \ldots, g_i^{(t_i)}$ be the elements of $GF(p_i^{n_i})$, $1 \le i \le m$, where $t_i = p_i^{n_i} - 1$. Let $n(s)$ be as in (4.1). Then,

$$\gamma_j = [g_1^{(j)}, g_2^{(j)}, \ldots, g_m^{(j)}], \ 0 \le j \le n(s), \tag{4.2}$$

possesses inverses and so does $\gamma_i - \gamma_j$ for $i \ne j$. Let us label the points γ in such a way that $0 = \gamma_0 = (0, 0, \ldots, 0)$ and the next $n(s)$ elements are given by (4.2). Form the $n(s)$ arrays L_j, whose (u, v)th element is filled by the element $\gamma_u + \gamma_j \gamma_v$ for $1 \le j \le n(s)$; $0 \le u, v \le s - 1$. Then, for a given j, L_j is a Latin square and $L_1, L_2, \ldots L_{n(s)}$ form a set of MOLS of order s (for a proof of this assertion, see e.g., Mann (1949, p. 140)).

Parker (1959a) showed that the MacNeish conjecture (also called the MacNeish-Mann conjecture) is false. Note that had the MacNeish-Mann conjecture been true, it would have shown the truth of Euler's conjecture as, by the MacNeish-Mann conjecture, $N(s) = 1$ if $s \equiv 2 \pmod 4$. Though the result of Parker cast doubts on the truth of Euler's conjecture, it did not show its falsity. In order to describe the result of Parker, we recall the definitions of balanced incomplete block (BIB) designs and, Mersenne and Fermat primes.

Let $\mathcal{V}$ be a finite set of v objects (or, *treatments*, using the terminology of statistical design of experiments) and $\mathcal{B}$, a collection of k-subsets of $\mathcal{V}$, where $2 \le k < v$; these subsets are called *blocks*. The pair $(\mathcal{V}, \mathcal{B})$ is a *balanced incomplete block* (BIB) design if (i) every treatment appears in r blocks and (ii) each pair of treatments occurs together in λ blocks. If $|\mathcal{B}| = b$, where $|\cdot|$ denotes the cardinality of a set, then the integers v, b, r, k, λ are called the parameters of a BIB design. A set of necessary (but not sufficient) conditions for the existence of a BIB design is

$$vr = bk, \ \lambda(v - 1) = r(k - 1), \ b \ge v.$$

In view of the first two indentities, one can write the parameters of a BIB design in terms of three independent parameters, say v, k and λ. Henceforth, we shall always denote a BIB design by the triple (v, k, λ). A BIB design (v, k, λ) is *symmetric* if $v = b$ and is said to be *resolvable* if its blocks can be partitioned into r sets, each set containing b/r blocks, such that each treatment appears exactly once in each set.

A *Mersenne prime* is a prime number of the form $2^p - 1 = M_p$ (say). For instance, $M_2 = 3, M_3 = 7, M_5 = 31$. If $2^p - 1$ is a prime, then so is p. However, the converse is not true; for example, $M_{11} = 2047 = (23)(89)$ is *not* a prime. A *Fermat prime* is a Fermat number $F_n = 2^{2^n} + 1$ which is a prime. For instance, $F_0 = 3, F_1 = 5, F_2 = 17$. No Fermat primes are known for $n > 4$.

Parker (1959a) proved the following result.

Theorem 4.1. *If there exists a BIB design* (v, k, λ) *with* $\lambda = 1$ *and* k *is the order of a finite projective plane, then there exists a set of* $k - 2$ *MOLS of order* v.

Theorem 4.1 was specialized and slightly strengthened by Parker (1959a) to yield the following result.

Theorem 4.2. *If* m *is a Mersenne prime larger than 3, or* $m + 1$ *is a Fermat prime larger than 3, then there exists a set of* m *MOLS of order* $m^2 + m + 1$.

The first case where Theorem 4.2 applies is $m = 4$, which yields a set of 4 MOLS of order $21(= 4^2 + 4 + 1)$. Note that for all orders covered by Theorem 4.2, $m \equiv 1 \pmod 3$ and $m^2 + m + 1 \equiv 3 \pmod 9$. Therefore, the construction of MacNeish produces only $n(s) = 2$ orthogonal Latin squares. We now know that there exist 5 MOLS of order $s = 21$ (A. V. Nazarok, 1991).

5. Falsity of Euler's Conjecture

The existence of an *Eulerian square* of order s defined in Section 2 is clearly equivalent to that of a pair of orthogonal Latin squares of order s. Eulerian squares are also known by the name *graeco-latin squares* in the statistical literature as, traditionally, one of the squares involved was written using Latin alphabets and the other, using Greek alphabets. In 1779, Euler started looking at the problem of finding Eulerian squares of every order. In fact, in his 1779 paper (which was published in 1782), Euler was able to construct an Eulerian square of every order s, where s is (i) either an odd integer or, (ii) a multiple of 4. Thus, the existence of Eulerian squares of all orders s where $s \equiv 0, 1$, or $3 \pmod 4$ was settled by Euler in 1782. The only case not settled till then was for orders $s \equiv 2 \pmod 4$. This brings us to the problem of 36 officers, stated below.

"There are 36 army officers, 6 from each rank and 6 from each regiment. Is it possible to arrange these 36 officers in a 6×6 square arrangement such that each rank and each regiment shows up in each row and each column?"

How did this problem arise in the first place? Folklore has the following 'explanation':

"It appears that the Emperor was to visit a garrison town in which six regiments were quartered and the commandant took into his head to arrange 36 officers in a square, one of each rank from each regiment, so that, whichever

8

row or column the Emperor walked along, he would meet one officer of each of the six ranks and one from each of the six regiments".

The commandant, of course, had set himself an impossible task as, the solution to the problem is provided by a 6×6 Eulerian square, which was later shown to be non-existent. Euler (1782) himself could not find an Eulerian square of order 6; he proceeded to 'show' the non-existence of such a square using an argument that is not entirely correct in method but correct in its conclusion. Having failed to construct an Eulerian square of order 6, Euler went on to make the following conjecture.

Euler's Conjecture: *No Eulerian square of order $s \equiv 2 \pmod 4$ exists.*

G. Tarry in 1900, by an exhaustive and laborious search showed the impossibility when $s = 6$. A shorter proof of the non-existence of an Eulerian square of order 6, based on coding theory was given by D. R. Stinson (1984). J. Peterson (1901) and P. Wernicke (1910) made errorneous attempts to prove Euler's conjecture as did MacNeish (1922). The arguments used by Peterson and MacNeish were shown to be false by F. W. Levi (1942) and the falsity of Wernicke's argument was shown by MacNeish (1921). Two leading statisticians, R. A. Fisher and F. Yates, in 1934 published a list of all possible Latin squares of order 6 and concluded as below:

"Euler's conclusion that no Greaco-Latin 6×6 square exists is easily verified from the 12 types of 6×6 Latin squares exemplified in this paper".

The first result casting serious doubts on the truth of Euler's conjecture is due to Bose and Shrikhande (1959) who were able to construct an Eulerian square of order $s = 22$. In their construction, Bose and Shrikhande (1959) used a class of designs called *pairwise balanced* designs, which may be viewed as a generalization of BIB designs. Let $\mathcal{V}$ be a finite set of v treatments and consider b blocks (subsets of $\mathcal{V}$), which are possibly of different sizes. These blocks form a pairwise balanced design of index unity and type $(v; k_1, k_2, \ldots, k_m)$ if each block contains either k_1 or k_2 or, $\ldots, k_m$ treatments and every pair of distinct treatments occurs exactly in one block, where for $1 \leq i \leq m$, $k_i \leq v$, $k_i \neq k_j$. Bose and Shrikhande (1959) proved the following result.

Lemma 5.1. *Suppose there exists a set S of $q - 1$ MOLS of order k, then we can construct a $q \times k(k - 1)$ matrix P, with entries $1, 2, \ldots, k$, such that any ordered pair $\binom{i}{j}$, $i \neq j$, occurs as a column exactly once in any two-rowed submatrix of P.*

Proof. Without loss of generality, let the first row of each Latin square in the set S have symbols $1, 2, \ldots, k$, in that order. Prefix the set S by a $k \times k$ array containing the symbol i in each position of the ith column. If the elements of each square are written in a single row such that the symbol in the ith row and jth column occupies the nth position in the row, where $n = k(i-1) + j$, then one can display these squares in the form of an orthogonal array $OA(k^2, q, k, 2)$. By deleting the first k columns, we get the desired matrix P. □

Let γ be a column of k distinct treatments $t_1, t_2, \ldots, t_k$ in that order. Let $P(\gamma)$ denote the $q \times k(k-1)$ matrix obtained by replacing the symbol i in P, by the treatment t_i occupying the ith position in γ, $1 \le i \le k$. Clearly every treatment occurs exactly $k-1$ times in every row of $P(\gamma)$ and any ordered pair $\binom{t_i}{t_j}$ occurs as a column exactly once in every two-rowed submatrix of $P(\gamma)$. We are now in a position to state the main result of Bose and Shrikhande (1959).

Theorem 5.1. *Let there exist a pairwise balanced design of index unity and type $(v; k_1, \ldots, k_m)$ and suppose there exist $q_i - 1$ MOLS of order k_i. If $q = min\{q_1, \ldots, q_m\}$, then there exist $q - 2$ MOLS of order v.*

Proof. Let the treatments of the pairwise balanced design be $t_1, \ldots, t_v$ and let the blocks of the design, written as columns which contain k_i treatments be denoted by $\gamma_{i1}, \ldots, \gamma_{ib_i}$, where for $1 \le i \le m$, b_i is the number of blocks of size k_i in the pairwise balanced design.

Let P_i be the matrix of order $q_i \times k(k-1)$ defined in Lemma 5.1, the elements of P_i being the symbols $1, 2, \ldots, k$. Let $C_{ij} = P_i(\gamma_{ij})$ be the matrix obtained from P_i and γ_{ij}. Suppose C_{ij}^* is obtained from C_{ij} by retaining any q rows. Define the $q \times v(v-1)$ matrix

$$C^* = [C_{11}^*, C_{12}^*, \ldots, C_{1b_1}^*, \ldots, C_{m1}^*, C_{m2}^*, \ldots, C_{mb_m}^*].$$

Let C_0^* be a $q \times v$ matrix whose ith column contains t_i in every position, $1 \le i \le v$. Then, $[C_0^*, C^*]$ is an orthogonal array $OA(v^2, q, v, 2)$. Using any two rows of this orthogonal array to coordinatize, we get $q - 2$ MOLS of order v. □

As an illustration of Theorem 5.1, consider a BIB design $(15, 3, 1)$, which has $b = 35$ blocks and each treatment is replicated $r = 7$ times. A resolvable solution for this design exists. Consider this resolvable BIB design and to each block of a replication, add a new treatment θ_i, $1 \le i \le 7$. Next, add a new block containing the treatments $\theta_1, \theta_2, \ldots, \theta_7$. This process gives us a pairwise balanced design of index unity and type $(22; 4, 7)$. Since there exist $q_1 = 3$ MOLS of order 4 and $q_2 = 6$ MOLS of order 7, an application of Theorem 5.1 shows that there exists a pair of orthogonal Latin squares of order 22, or equivalently, an Eulerian square of order 22.

In the same year, Parker (1959b) proved the following result.

Theorem 5.2. *There exists an Eulerian square of order $s = (3q - 1)/2$, where $q = 3 \pmod 4$ is a prime or a prime power > 3.*

Parker's method fails for $q = 3$. For $q = 7$, one obtains a pair of orthogonal Latin squares of order 10, or an Eulerian square of order 10. This square is shown below.

00	47	18	76	29	93	85	34	61	52
86	11	57	28	70	39	94	45	02	63
95	80	22	67	38	71	49	56	13	04
59	96	81	33	07	48	72	60	24	15
73	69	90	82	44	17	58	01	35	26
68	74	09	91	83	55	27	12	46	30
37	08	75	19	92	84	66	23	50	41
14	25	36	40	51	62	03	77	88	99
21	32	43	54	65	06	10	89	97	78
42	53	64	05	16	20	31	98	79	87

Once Eulerian squares of orders 10 and 22 were found, more doubts about the validity of Euler's conjecture arose as both 10 and 22 are congruent to 2 mod 4. The result in Theorem 5.1, which may be viewed as a generalization of Theorem 4.1, was further strengthened by Bose and Shrikhande (1960), again using pairwise balanced designs of index unity. Consider a pairwise balanced design d of index unity and of the type $(v; k_1, k_2, \ldots, k_m)$ and as before, for $1 \leq i \leq m$, let b_i be the number of blocks in d which are of size k_i. Let d_i, $1 \leq i \leq m$, be the subdesign of d consisting of the b_i blocks of size k_i each. Then, d_i, $1 \leq i \leq m$, is called the ith *equiblock component* of d. A subset of blocks belonging to any equiblock component d_i is said to be of type I if every treatment occurs in the subset exactly k_i times. A subset of blocks belonging to any equiblock component d_i is said to be of type II if every treatment occurs in the subset exactly once. The component d_i is said to be *separable* if the blocks can be divided into subsets of type I or type II. The design d is called separable if each equiblock component d_i is separable. Bose and Shrikhande (1960) proved the following result.

Theorem 5.3. *Let there exist a pairwise balanced design d of index unity and type $(v; k_1, \ldots, k_m)$ and suppose there exist $q_i - 1$ MOLS of order k_i. If*

$$q = min\{q_1, q_2, \ldots, q_m\}$$

then there exist at least $q - 2$ MOLS of order v. Furthermore, if d is separable, then the number of MOLS of order v is at least $q - 1$.

Numerous applications of Theorem 5.3 were made by Bose and Shrikhande (1960) to obtain sets of MOLS. In particular, they proved the following result.

Theorem 5.4. *There exist at least two orthogonal Latin squares of order $s = 36m + 22$, where $m \geq 0$ is an integer.*

Theorem 5.4 shows the falsity of the Euler's conjecture for infinitely many values of $s \equiv 2 \pmod 4$, $s \geq 22$.

That the Euler's conjecture is false for *all* orders $s = 4t + 2$, $t > 1$ was shown by Bose, Shrikhande and Parker (1960). Their method consisted of finding an appropriate pairwise balanced design of index unity. Recall the definition of equiblock components of a pairwise balanced design of index unity. A set of equiblock components $d_1, d_2, \ldots, d_l$, $l < m$, is said to be a *clear* set if any pair of blocks among the $\sum_{i=1}^{l} b_i$ blocks comprising $d_1, \ldots, d_l$ are disjoint. The main result of Bose, Shrikhande and Parker (1960) can now be stated.

Theorem 5.5. *Let there exist a pairwise balanced design d of index unity and type $(v; k_1, k_2, \ldots, k_m)$ such that the equiblock components $d_1, d_2, \ldots, d_l$, $l < m$, is a clear set. If there exists a set of $q_i - 1$ MOLS of order k_i, $1 \leq i \leq m$, and if $q^* = min\{q_1 + 1, \ldots, q_l + 1, q_{l+1}, \ldots, q_m\}$, then there exist at least $q^* - 2$ MOLS of order v.*

Bose, Shrikhande and Parker (1960) then gave methods of construction of pairwise balanced designs of the type demanded in Theorem 5.5 using BIB designs with $\lambda = 1$, resolvable BIB designs and *group divisible* (GD) designs. An arrangement of $v = lm$ treatments and b blocks, each containing k distinct treatments, $(2 \leq k < v)$, is said to be a GD design if the treatments can be grouped into l groups of m treatments each, such that any two treatments of the same group appear together in λ_1 blocks and any pair of treatments belonging to different groups appear together in λ_2 blocks. In the context of orthogonal Latin squares, GD designs with $\lambda_1 = 0, \lambda_2 = 1$ play a special role as, these can be used to obtain pairwise balanced designs required in Theorem 5.5. A group divisible design is denoted by $GD(v; k, m; \lambda_1, \lambda_2)$. Let r denote the common replication of a treatment in a GD design. GD designs can be classified into three sub-classes: (i) singular, if $r - \lambda_1 = 0$; (ii) semi-regular, if $r - \lambda_1 > 0$, $rk - v\lambda_2 = 0$; (iii) regular, if $r - \lambda_1 > 0$, $rk - v\lambda_2 > 0$.

We first consider an application of Theorem 5.5. Consider a BIB design with $\lambda = 1$, which we shall denote by $BIB(v, k)$. If we delete any three treatments, say $\alpha_1, \alpha_2, \alpha_3$ not occurring in the same block of a $BIB(v, k)$, we get a pairwise balanced design d of index unity and type $(v - 3; k, k - 1, k - 2)$. Since in a $BIB(v, k)$, no pair of blocks has more than one treatment in common, the three blocks of d obtained by deleting (α_1, α_2), (α_1, α_3) and (α_2, α_3) have no treatment in common and thus form a clear equiblock component. Hence, we get the following result.

Theorem 5.6. *The existence of a $BIB(v, k)$ implies that*

$$N(v - 3) \geq min\{N(k), N(k - 1), N(k - 2) + 1\} - 1.$$

12

As an application of Theorem 5.6, consider $BIB(v, k)$ designs with $k = 5$ and $v = 21, 25, 41, 45, 61, 65, 85, 125$, all of which do exist. By invoking Theorem 5.6, we see that there there exist at least two MOLS of each of the following orders: 18, 22, 38, 42, 58, 62, 82 and 122.

Using a resolvable $GD(km; k, m; 0, 1)$ design, Bose, Shrikhande and Parker (1960) proved the following result.

Theorem 5.7. *If $k \leq N(m) + 1$, then*
(i) $N(km + 1) \geq min\{N(k), N(k + 1), N(m) + 1\} - 1$,
(ii) $N(km + x) \geq min\{N(k), N(k + 1), N(m) + 1, N(x) + 1\} - 1$ if $1 < x < m$.

Towards proving the falsity of Euler's conjecture for all orders $s \equiv 2 \pmod{4}$, $s > 6$, Bose, Shrikhande and Parker (1960) first proved the following result.

Lemma 5.2. *There exist at least two MOLS of order $s \equiv 2 \pmod{4}$ if $6 < s \leq 726$.*

Proof. Bose, Shrikhande and Parker (1960) first showed that there exist at least two MOLS of order s, where $10 \leq s \leq 154$, $s \equiv 2 \pmod{4}$, by improving upon the lower bounds on $N(s)$ given by Bose and Shrikhande (1960). Any integer lying in the closed interval $I_i = [a_i, b_i]$ shown in the following table can be expressed in the form

$$s = 4m_i + x_i, \ 10 \leq x_i \leq c_i,$$

where m_i and c_i are as given in the following table (*cf.* Bose, Shrikhande and Parker, 1960).

i	$I_i[a_i, b_i]$	m_i	c_i
1	[158, 182]	37	34
2	[186, 218]	44	42
3	[222, 262]	53	50
4	[266, 310]	64	54
5	[314, 374]	76	70
6	[378, 454]	92	86
7	[458, 550]	112	102
8	[554, 662]	136	118
9	[666, 726]	164	70

It is easily seen that $N(m_i) \geq n(m_i) \geq 3$. Also, $N(x_i) \geq 2$, since $10 \leq x_i \leq c_i < 154$. If we take $k = 4$ in part (ii) of Theorem 5.7, the conditions $k \leq N(m_i) + 1$ and $1 < x_i < m_i$ are satisfied. Hence, if s lies in any of the closed

13

intervals I_i $(1 \le i \le 9)$, $N(s) \ge 2$. The proof is completed by noting that any $s \equiv 2 \pmod 4$ satisfying $154 < s \le 726$ lies in one of the intervals I_i. $\square$

The following is the final result due to Bose, Shrikhande and Parker (1960), proving the falsity of Euler's conjecture.

Theorem 5.8. *There exists at least two MOLS of side $s > 2$, $s \ne 6$.*

Proof. If $s \equiv 0 \pmod 4$ or s is odd, then $N(s) \ge 2$, as shown by Euler (1782). In view of Lemma 5.2, it thus suffices to prove the result for orders $s \equiv 2 \pmod 4$, $s \ge 730$. If s satisfies these conditions, one can write

$$s - 10 = 144g + 4u, \quad g \ge 5, \; 0 \le u \le 35,$$

so that $s = 4(36g) + 4u + 10$. In the prime power decomposition of $36g$, the least factor is greater than or equal to 4, and thus, $N(36g) \ge 3$. If in part (ii) of Theorem 5.7, we take $k = 4, m = 36g, x = 4u + 10$, then $k \le 1 + N(m)$. Also, $10 \le x \le 150, m \ge 180$. Hence $1 < x < m$ and $N(x) \ge 2$. It follows now that $N(s) \ge 2$. $\square$

We conclude this section by noting that using universal algebra, T. Evans (1982) showed that there are an infinite number of values of $s \equiv 2 \pmod 4$ for which there exists an Eulerian square of order $s > 6$.

6. Historical Notes

The literature on Latin squares is at least three centuries old, one of the earliest references being a monograph *Koo-Soo-Ryak* by Choi Seok-Jeong (1646–1715), who used orthogonal Latin squares of order 9 to construct a magic square and stated that he cannot find orthogonal Latin squares of order 10.

Recall that a (traditional) *magic square* of order $n \ge 2$ with magic constant $e = n(n^2 + 1)/2$ is an $n \times n$ matrix $A = (a_{ij})$ with entries $1, 2, \ldots, n^2$, such that:

(i) $\sum_{i=1}^{n} a_{ij} = e, 1 \le j \le n$,
(ii) $\sum_{j=1}^{n} a_{ij} = e, 1 \le i \le n$,
(iii) $\sum_{i=1}^{n} a_{ii} = e$, and,
(iv) $\sum_{i=1}^{n} a_{i,(n-i+1)} = e$.

As noted by I. Anderson, C. J. Colbourn, J. H. Dinitz and T. S. Griggs (2007), Latin square amulets go back to medieval Islam (c1200). A magic square of the famous Arab sufi, Ahmad ibn Ali ibn Yusuf al-Buni indicates the knowledge of a pair of orthogonal Latin squares of order 4. A new edition of J. Ozanam's four-volume treatise "Récréations mathématiques et physiques ...", published in 1723 had the following puzzle:

"There are 16 playing cards of four denominations, ace, king, queen and jack from each of the four suits, spade, heart, diamond and club. Is it possible to arrange these 16 cards in a 4×4 square such that each denomination and each suit appears in each row, each column and (additionally) on the two diagonals exactly once?"

It can be verified easily that the Eulerian square of order 4 given in Section 2 provides a solution to the above puzzle. Thus, there is evidence of the existence of Eulerian squares much before Euler!

Euler's interest in this area also probably originated from the connection of Eulerian squares to magic squares. Euler, in a paper entitled "De quadratis magicis" and read before the Academy of Sciences at St. Petersburg on October 17, 1776, constructed magic squares of orders 3, 4 and 5 from orthogonal Latin squares. He could not construct an Eulerian square of order 6 which prompted him to make his conjecture. For over a century, no progress was made on this problem, though it was not totally neglected by mathematicians of that time. In 1842, C. F. Gauss and H. C. Schumacher corresponded about a work of T. Clausen, who apparently established the impossibility of an Eulerian square when $s = 6$ and conjectured the impossibility when $s = 2$ (mod 4). This work unfortunately was never published!

In 1896, E. H. Moore published a very influential paper "Tactical Memoranda I–III" in the American Journal of Mathematics. In Memorandum II of this paper, Moore used finite fields of order s to construct a complete set of MOLS of order s, a result rediscovered subsequently by Bose (1938) and independently by Stevens (1939). Moore also proved that if there exist t MOLS of order m and of order n, then there exist t MOLS of order mn, a fact established later by MacNeish (1922) also. MacNeish, Bose, Stevens were all apparently unaware of the contribution of Moore.

The final results of Bose, Shrikhande and Parker on the falsity of Euler's conjecture for all orders $s \equiv 2$ (mod 4), $s > 6$, were announced in the annual meeting of the American Mathematical Society, held in New York during the last week of April, 1959. This major result was reported on the front page of the Sunday edition of the New York Times of April 26, 1959. Since then the trio Bose-Shrikhande-Parker became known as "Euler's Spoilers"!

Acknowledgment

This work was supported by the Indian National Science Academy under the Senior Scientist scheme of the Academy. The support is gratefully acknowledged.

15

References

Anderson, I., C. J. Colbourn, J. H. Dinitz and T. S. Griggs (2007). Design theory: Antiquity to 1950. In: *Handbook of Combinatorial Designs*, 2nd. ed. (C. J. Colbourn and J. H. Dinitz, Eds.). New York: Chapman and Hall/CRC, pp. 11–22.

Bose, R. C. (1938). On the application of the properties of Galois fields to the problem of construction of hyper-Greaco-latin squares. *Sankhyā* **3**, 323–338.

Bose, R. C. and S. S. Shrikhande (1959). On the falsity of Euler's conjecture about the non-existence of two orthogonal latin squares of order 4t+2. *Proc. Natl. Acad. Sci. U. S. A.* **45**, 734–737.

Bose, R. C. and S. S. Shrikhande (1960). On the construction of sets of mutually orthogonal latin squares and the falsity of a conjecture of Euler. *Trans. Amer. Math. Soc.* **95**, 191–209.

Bose, R. C., S. S. Shrikhande and E. T. Parker (1960). Further results on the constructon of mutually orthogonal latin squares and the falsity of Euler's conjecture. *Can. J. Math.* **12**, 189–203.

Bruck, R. H. and H. J. Ryser (1949). The nonexistence of certain finite projective planes. *Can. J. Math.* **1**, 88–93.

Chowla, S. and H. J. Ryser (1950). Combinatorial problems. *Can. J. Math.* **2**, 93–99.

Dénes, J. and A. D. Keedwell (1974). *Latin Squares and Their Applications*. New York: Academic Press.

Dénes, J. and A. D. Keedwell (eds.) (1991). *Latin Squares: New Developments in the Theory and Applications*. Amsterdam: North-Holland. Annals of Discrete Mathematics **46**.

Euler, L. (1782). Recherches sur une nouvelle espece de quarrés magiques. *Verh. Zeeuw. Gen. Weten. Vlissengen* **9**, 85–239.

Evans, T. (1982). Universal algebra and Euler's officer problem. *Amer. Math. Monthly* **86**, 466–479.

Fisher, R. A. and F. Yates (1934). The 6×6 Latin squares. *J. Cambridge Phil. Soc.* **30**, 492–507.

Lam, C. W. H., L. H. Thiel and S. Swiercz (1989). The nonexistence of finite projective planes of order 10. *Can. J. Math.* **41**, 1117–1123.

Laywine, C. F. and G. L. Mullen (1998). *Discrete Mathematics Using Latin Squares*. New York: Wiley.

Levi, F. W. (1942). *Finite Geometrical Systems*. University of Calcutta.

MacNeish, H. F. (1921). Das problem der 36 offiziere. *Ber. Deuts. Mat. Ver.* **30**, 151–153.

MacNeish, H. F. (1922). Euler squares. *Ann. Math.* **23**, 221–227.

Mann, H. B. (1942). The construction of orthogonal latin squares. *Ann. Math. Statist.* **13**, 418–423.

Mann, H. B. (1949). *Analysis and Design of Experiments*. New York: Dover.

Moore, E. H. (1896). Tactical memoranda I–III. *Amer. J. Math.* **18**, 264–303.

Nazarok, A. V. (1991). Five pairwise orthogonal latin squares of order 21. *Issled. Oper. ASU*, 54–56.

Ozanam, J. (1723). *Récréations Mathématiques et Physiques, qui contiennent Plusieurs Problémes utiles & agréables, d'Arithmetique, de Geometrie, d'Optique, de Gnomonique, de Cosmographie, de Mécanique, de Pyrotechnie, & de Physique.* 4 Vols. Paris: Jombert (updated edition).

Parker, E. T. (1959a). Construction of some sets of pairwise orthogonal Latin squares. *Proc. Amer. Math. Soc.* **10**, 946–951.

Parker, E. T. (1959b). Orthogonal latin squares. *Proc. Natl. Acad. Sci. U. S. A.* **45**, 859–862.

Peterson, J. (1901). Les 36 officieurs. *Ann. Math.* **1**, 413–427.

Shrikhande, S. S. (1950). The impossibility of certain symmetrical balanced incomplete block designs. *Ann. Math. Statist.* **21**, 106–111.

Stevens, W. L. (1939). The completely orthogonalised Latin squares. *Ann. Eugen.* **9**, 82–93.

Stinson, D. R. (1984). A short proof of the nonexistence of a pair of orthogonal Latin squares of order six. *J. Combin. Theor. Ser. A* **36**, 373–376.

Tarry, G. (1900). Le probléme des 36 officers. *Comptes Rendus de l'Association Francaise pour l'Avancement des Sciences: Série de mathématiques, astronomie, géodésie et mécanique* **29**, 170–203.

Wernicke, P. (1910). Das problem der 36 offiziere. *Deutsche Math.-Ver.* **19**, 264–267.

Cramér-Rao Lower Bound and Information Geometry

Frank Nielsen[*]

1. Introduction and Historical Background

This article focuses on an important piece of work of the world renowned Indian statistician, Calyampudi Radhakrishna Rao. In 1945, C. R. Rao (25 years old then) published a pathbreaking paper [43], which had a profound impact on subsequent statistical research. Roughly speaking, Rao obtained a lower bound to the variance of an estimator. The importance of this work can be gauged, for instance, by the fact that it has been reprinted in the volume *Breakthroughs in Statistics: Foundations and Basic Theory* [32]. There have been two major impacts of this work:

- First, it answers a fundamental question statisticians have always been interested in, namely, how good can a statistical estimator be? Is there a fundamental limit when estimating statistical parameters?

- Second, it opens up a novel paradigm by introducing differential geometric modeling ideas to the field of Statistics. In recent years, this contribution has led to the birth of a flourishing field of *Information Geometry* [6].

It is interesting to note that H. Cramér [20] (1893-1985) also dealt with the same problem in his classic book *Mathematical Methods of Statistics*, published in 1946, more or less at the same time Rao's work was published. The result is widely acknowledged nowadays as the Cramér-Rao lower bound (CRLB). The lower bound was also reported independently[1] in the work of M. Fréchet [27] (uniparameter case) and G. Darmois [22] (multi-parameter case). The Fréchet-Darmois work were both published in French, somewhat limiting its international scientific exposure. Thus the lower bound is also sometimes called the Cramér-Rao-Fréchet-Darmois lower bound.

---

[*]Sony Computer Science Laboratories Inc., Japan, École Polytechnique, LIX, France
[1]The author thanks F. Barbaresco for bringing the historical references to his attention.

This review article is organized as follows: Section 2 introduces the two fundamental contributions in C. R. Rao's paper:

- The Cramér-Rao lower bound (CRLB), and

- The Fisher-Rao Riemannian geometry.

Section 3 concisely explains how information geometry has since evolved into a full-fledged discipline. Finally, Section 4 concludes this review by discussing further perspectives of information geometry and hinting at the future challenges.

2. Two Key Contributions to Statistics

To begin with, we describe the two key contributions of Rao [43], namely a lower bound to the variance of an estimator and Rao's Riemannian information geometry.

2.1. Rao's Lower Bound for Statistical Estimators. For a fixed integer $n \geq 2$, let $\{X_1, ..., X_n\}$ be a random sample of size n on a random variable X which has a probability density function (pdf) (or, probability mass function (pmf)) $p(x)$. Suppose the unknown distribution $p(x)$ belongs to a parameterized family $\mathcal{F}$ of distributions

$$\mathcal{F} = \{p_\theta(x) \mid \theta \in \Theta\},$$

where θ is a parameter vector belonging to the parameter space Θ. For example, $\mathcal{F}$ can be chosen as the family $\mathcal{F}_{\text{Gaussian}}$ of all normal distributions with parameters $\theta = (\mu, \sigma)$ (with $\theta \in \Theta = \mathbb{R} \times \mathbb{R}^+$):

The unknown distribution $p(x) = p_{\theta^*}(x) \in \mathcal{F}$ is identified by a unique parameter $\theta^* \in \Theta$. One of the major problems in Statistics is to build an "estimator" of θ^* on the basis of the sample observations $\{X_1, \ldots, X_n\}$.

There are various estimation procedures available in the literature, e.g., the method of moments and the method of maximum likelihood; for a more comprehensive account on estimation theory, see e.g., [33]. From a given sample of fixed size n, one can get several estimators of the same parameter. A natural question then is: which estimator should one use and how their performance compare to each other. This is related precisely with C. R. Rao's first contribution in his seminal paper [43]. Rao addresses the following question:

What is the *accuracy attainable* in the estimation of statistical parameters?

Before proceeding further, it is important to make some observations on the notion of *likelihood*, introduced by Sir R. A. Fisher [26]. Let $\{X_1, \ldots, X_n\}$ be a random vector with pdf (or, pmf) $p_\theta(x_1, \ldots, x_n)$, $\theta \in \Theta$, where for $1 \le i \le n$, x_i is a realization of X_i. The function

$$L(\theta; x_1, \ldots, x_n) = p_\theta(x_1, \ldots, x_n),$$

considered as a function of θ, is called the likelihood function. If $X_1, \ldots, X_n$ are independent and identically distributed random variables with pdf (or, pmf) $p_\theta(x)$ (for instance, if $X_1, \ldots, X_n$ is a random sample from $p_\theta(x)$), the likelihood function is

$$L(\theta; x_1, \ldots, x_n) = \prod_{i=1}^{n} p_\theta(x_i).$$

The method of maximum likelihood estimation consists of choosing an estimator of θ, say $\hat\theta$ that maximizes $L(\theta; x_1, \ldots, x_n)$. If such a $\hat\theta$ exists, we call it a *maximum likelihood estimator* (MLE) of θ. Maximizing the likelihood function is mathematically equivalent to maximizing the log-likelihood function since the logarithm function is a strictly increasing function. The log-likelihood function is usually simpler to optimize. We shall write $l(x, \theta)$ to denote the log-likelihood function with $x = (x_1, \ldots, x_n)$. Finally, we recall the definition of an *unbiased* estimator. Let $\{p_\theta, \theta \in \Theta\}$ be a set of probability distribution functions. An estimator T is said to be an unbiased estimator of θ if the expectation of T,

$$E_\theta(T) = \theta, \quad \text{for all } \theta \in \Theta.$$

Consider probability distributions with pdf (or, pmf) satisfying the following *regularity conditions*:

- The support $\{x \mid p_\theta(x) > 0\}$ is identical for all distributions (and thus does not depend on θ),

- $\int p_\theta(x)dx$ can be differentiated under the integral sign with respect to θ,

- The gradient $\nabla_\theta p_\theta(x)$ exists.

We are now ready to state C. R. Rao's fundamental limit of estimators.

2.1.1. Rao's Lower Bound: Single Parameter Case. Let us first consider the case of uni-parameter distributions like Poisson distributions with mean parameter λ. These families are also called order-1 families of probabilities. The C. R. Rao lower bound in the case of uni-parameter distributions can be stated now.

Theorem 1 (Rao lower bound (RLB)). *Suppose the regularity conditions stated above hold. Then the variance of any unbiased estimator $\hat\theta$, based on an independent and identically distributed (IID) random sample of size n, is bounded* *below by* $\frac{1}{nI(\theta^*)}$, *where* $I(\theta)$ *denotes the Fisher information in a single observation, defined as*

$$I(\theta) = -E_\theta \left[\frac{\mathrm{d}^2 l(x;\theta)}{\mathrm{d}\theta^2} \right] = \int -\frac{\mathrm{d}^2 l(x;\theta)}{\mathrm{d}\theta^2} p_\theta(x)\mathrm{d}x.$$

As an illustration, consider the family of Poisson distributions with parameter $\theta = \lambda$. One can check that the regularity conditions hold. For a Poisson distribution with parameter λ, $l(x;\lambda) = -\lambda + \log \frac{\lambda^x}{x!}$ and thus,

$$l'(x;\lambda) = -1 + \frac{x}{\lambda},$$
$$l''(x;\lambda) = -\frac{x}{\lambda^2}.$$

The first derivative is technically called the *score function*. It follows that

$$I(\lambda) = -E_\lambda \left[\frac{\mathrm{d}^2 l(x;\lambda)}{\mathrm{d}\lambda^2} \right],$$
$$= \frac{1}{\lambda^2} E_\lambda[x] = \frac{1}{\lambda}$$

since $E[X] = \lambda$ for a random variable X following a Poisson distribution with parameter λ: $X \sim \mathrm{Poisson}(\lambda)$. What the RLB theorem states in plain words is that for any unbiased estimator $\hat\lambda$ based on an IID sample of size n of a Poisson distribution with parameter $\theta^* = \lambda^*$, the variance of $\hat\lambda$ cannot go below $\frac{1}{nI(\lambda^*)} = \lambda^*/n$.

The Fisher information, defined as the variance of the score, can be geometrically interpreted as the curvature of the log-likelihood function. When the curvature is low (log-likelihood curve is almost flat), we may expect some large amount of deviation from the optimal θ^*. But when the curvature is high (peaky log-likelihood), we rather expect a small amount of deviation from θ^*.

2.1.2. Rao's Lower Bound: Multi-parameter Case. For d-dimensional multi-parameter[2] distributions, the Fisher information matrix $I(\theta)$ is defined as the symmetric matrix with the following entries [6]:

$$[I(\theta)]_{ij} = E_\theta \left[\frac{\partial}{\partial\theta_i} \log p_\theta(x) \frac{\partial}{\partial\theta_j} \log p_\theta(x) \right], \tag{1}$$

$$= \int \left(\frac{\partial}{\partial\theta_i} \log p_\theta(x) \frac{\partial}{\partial\theta_j} \log p_\theta(x) \right) p_\theta(x)\mathrm{d}x. \tag{2}$$

---

[2]Multi-parameter distributions can be univariate like the 1D Gaussians $N(\mu,\sigma)$ or multivariate like the Dirichlet distributions or dD Gaussians.

Provided certain regularity conditions are met (see [6], section 2.2), the Fisher information matrix can be written *equivalently* as:

$$[I(\theta)]_{ij} = -E_\theta \left[\frac{\partial^2}{\partial\theta_i\partial\theta_j} \log p_\theta(x) \right],$$

or as:

$$[I(\theta)]_{ij} = 4 \int_{x\in\mathcal{X}} \frac{\partial}{\partial\theta_i} \sqrt{p_\theta(x)} \frac{\partial}{\partial\theta_j} \sqrt{p_\theta(x)} \mathrm{d}x.$$

In the case of multi-parameter distributions, the lower bound on the accuracy of unbiased estimators can be extended using the Löwner partial ordering on matrices defined by $A \succeq B \Leftrightarrow A - B \succeq 0$, where $M \succeq 0$ means M is positive semidefinite [11] (We similarly write $M \succ 0$ to indicate that M is positive definite).

The Fisher information matrix is always positive semi-definite [33]. It can be shown that the Fisher information matrix of regular probability distributions is positive definite, and therefore always invertible. Theorem 1 on the lower bound on the inaccuracy extends to the multi-parameter setting as follows:

Theorem 2 (Multi-parameter Rao lower bound (RLB)). *Let θ be a vector-valued parameter. Then for an unbiased estimator $\hat\theta$ of θ^* based on a IID random sample of n observations, one has $V[\hat\theta] \succeq n^{-1}I^{-1}(\theta^*)$, where $V[\hat\theta]$ now denotes the variance-covariance matrix of $\hat\theta$ and $I^{-1}(\theta^*)$ denotes the inverse of the Fisher information matrix evaluated at the optimal parameter θ^*.*

As an example, consider a IID random sample of size n from a normal population $N(\mu^*, \sigma^{*2})$, so that $\theta^* = (\mu^*, \sigma^{*2})$. One can then verify that the Fisher information matrix of a normal distribution $N(\mu, \sigma^2)$ is given by

$$I(\theta) = \begin{bmatrix} \frac{1}{\sigma^2} & 0 \\ 0 & \frac{1}{2\sigma^4} \end{bmatrix}.$$

Therefore,

$$V[\hat\theta] \succeq n^{-1}I(\theta^*)^{-1} = \begin{bmatrix} n^{-1}\sigma^{*2} & 0 \\ 0 & 2n^{-1}\sigma^{*4} \end{bmatrix}.$$

There has been a continuous flow of research along the lines of the CRLB, including the case where the Fisher information matrix is singular (positive semidefinite, e.g. in statistical mixture models). We refer the reader to the book of Watanabe [47] for a modern algebraic treatment of degeneracies in statistical learning theory.

2.2. Rao's Riemannian Information Geometry.

What further makes C. R. Rao's 1945 paper a truly impressive milestone in the development of Statistics is the introduction of differential geometric methods for modeling population spaces using the Fisher information matrix. Let us review the framework that literally opened up the field of information geometry [6].

Rao [43] introduced the notions of the Riemannian Fisher information metric and geodesic distance to the Statisticians. This differential geometrization of Statistics gave birth to what is known now as the field of *information geometry* [6]. Although there were already precursor geometric work [35, 12, 36] linking geometry to statistics by the Indian community (Professors Mahalanobis and Bhattacharyya), none of them studied the differential concepts and made the connection with the Fisher information matrix. C. R. Rao is again a pioneer in offering Statisticians the geometric lens.

2.2.1. The Population Space. Consider a family of parametric probability distribution $p_\theta(x)$ with $x \in \mathbb{R}^d$ and $\theta \in \mathbb{R}^D$ denoting the D-dimensional parameters of distributions (order of the probability family). The population parameter space is the space

$$\Theta = \left\{ \theta \in \mathbb{R}^D \middle| \int p_\theta(x) \mathrm{d}x = 1 \right\}.$$

A given distribution $p_\theta(x)$ is interpreted as a corresponding point indexed by $\theta \in \mathbb{R}^D$. θ also encodes a coordinate system to identify probability models: $\theta \leftrightarrow p_\theta(x)$.

Consider now two infinitesimally close points θ and $\theta + \mathrm{d}\theta$. Their probability densities differ by their first order differentials: $\mathrm{d}p(\theta)$. The distribution of $\mathrm{d}p$ over all the support aggregates the consequences of replacing θ by $\theta + \mathrm{d}\theta$. Rao's revolutionary idea was to consider the *relative discrepancy* $\frac{\mathrm{d}p}{p}$ and to take the variance of this difference distribution to define the following quadratic differential form:

$$
\begin{aligned}
\mathrm{d}s^2(\theta) &= \sum_{i=1}^{D} \sum_{j=1}^{D} g_{ij}(\theta) \mathrm{d}\theta_i \mathrm{d}\theta_j, \\
&= (\nabla\theta)^\top G(\theta) \nabla\theta,
\end{aligned}
$$

with the matrix entries of $G(\theta) = [g_{ij}(\theta)]$ as

$$g_{ij}(\theta) = E_\theta \left[\frac{1}{p(\theta)} \frac{\partial p}{\partial \theta_i} \frac{1}{p(\theta)} \frac{\partial p}{\partial \theta_j} \right] = g_{ji}(\theta).$$

In differential geometry, we often use the symbol ∂_i as a shortcut to $\frac{\partial}{\partial \theta_i}$.

23

The elements $g_{ij}(\theta)$ form the quadratic differential form defining the elementary length of Riemannian geometry. The matrix $G(\theta) = [g_{ij}(\theta)] \succ 0$ is positive definite and turns out to be equivalent to the *Fisher information matrix*: $G(\theta) = I(\theta)$. The information matrix is invariant to monotonous transformations of the parameter space [43] and makes it a good candidate for a Riemannian metric.

We shall discuss later more on the concepts of invariance in statistical manifolds [18, 38].

In [43], Rao proposed a novel versatile notion of statistical distance induced by the Riemannian geometry beyond the traditional Mahalanobis D-squared distance [35] and the Bhattacharyya distance [12]. The Mahalanobis D-squared distance [35] of a vector x to a group of vectors with covariance matrix Σ and mean μ is defined originally as

$$D_\Sigma^2(x, \mu) = (x - \mu)^\top \Sigma^{-1} (x - \mu).$$

The generic Mahalanobis distance $D_M(p, q) = \sqrt{(p - q)^\top M (p - q)}$ (with M positive definite) generalizes the Euclidean distance (M chosen as the identity matrix).

The Bhattacharyya distance [12] between two distributions indexed by parameters θ_1 and θ_2 is defined by

$$B(\theta_1, \theta_2) = - \log \int_{x \in \mathcal{X}} \sqrt{p_{\theta_1}(x) p_{\theta_2}(x)} \mathrm{d}x.$$

Although the Mahalanobis distance D_M is a metric (satisfying the triangle inequality and symmetry), the Bhattacharyya distance fails the triangle inequality. Nevertheless, it can be used to define the Hellinger metric distance

$$H(\theta_1, \theta_2) = \sqrt{1 - e^{-B(\theta_1, \theta_2)}}.$$

2.2.2. Rao's Distance: Riemannian Distance Between Two Populations.
Let P_1 and P_2 be two points of the population space corresponding to the distributions with respective parameters θ_1 and θ_2. In Riemannian geometry, the geodesics are the *shortest paths*. For example, the geodesics on the sphere are the arcs of great circles. The statistical distance between the two populations is defined by integrating the infinitesimal element lengths $\mathrm{d}s$ along the geodesic linking P_1 and P_2. Equipped with the Fisher information matrix tensor $I(\theta)$, the *Rao distance* $D(\cdot, \cdot)$ between two distributions on a statistical manifold can be calculated from the geodesic length as follows:

$$D(p_{\theta_1}(x), p_{\theta_2}(x)) = \min_{\substack{\theta(t) \\ \theta(0)=\theta_1, \theta(1)=\theta_2}} \int_0^1 \left(\sqrt{(\nabla \theta)^\top I(\theta) \nabla \theta} \right) \mathrm{d}t$$

Therefore we need to calculate explicitly the geodesic linking $p_{\theta_1}(x)$ to $p_{\theta_2}(x)$ to compute Rao's distance. This is done by solving the following second order ordinary differential equation (ODE) [6]:

$$g_{ki}\ddot{\theta}_i + \Gamma_{k,ij}\dot{\theta}_i\dot{\theta}_j = 0,$$

where Einstein summation [6] convention has been used to simplify the mathematical writing by removing the leading sum symbols. The coefficients $\Gamma_{k,ij}$ are the Christoffel symbols of the first kind defined by:

$$\Gamma_{k,ij} = \frac{1}{2}\left(\frac{\partial g_{ik}}{\partial \theta_j} + \frac{\partial g_{kj}}{\partial \theta_i} - \frac{\partial g_{ij}}{\partial \theta_k}\right).$$

For a parametric statistical manifold with D parameters, there are D^3 Christoffel symbols. In practice, it is difficult to explicitly compute the geodesics of the Fisher-Rao geometry of arbitrary models, and one needs to perform a gradient descent to find a local solution for the geodesics [41]. This is a drawback of the Rao's distance as it has to be checked manually whether the integral admits a closed-form expression or not.

To give an example of the Rao distance, consider the smooth manifold of univariate normal distributions, indexed by the $\theta = (\mu, \sigma)$ coordinate system. The Fisher information matrix is

$$I(\theta) = \begin{bmatrix} \frac{1}{\sigma^2} & 0 \\ 0 & \frac{2}{\sigma^2} \end{bmatrix} \succ 0. \tag{3}$$

The infinitesimal element length is:

$$\begin{aligned} ds^2 &= (\nabla\theta)^\top I(\theta)\nabla\theta, \\ &= \frac{d\mu^2}{\sigma^2} + \frac{2d\sigma^2}{\sigma^2}. \end{aligned}$$

After the minimization of the path length integral, the Rao distance between two normal distributions [43, 8] $\theta_1 = (\mu_1, \sigma_1)$ and $\theta_2 = (\mu_2, \sigma_2)$ is given by:

$$D(\theta_1, \theta_2) = \begin{cases} \sqrt{2}\log\frac{\sigma_2}{\sigma_1} & \text{if } \mu_1 = \mu_2, \\ \frac{|\mu_1 - \mu_2|}{\sigma} & \text{if } \sigma_1 = \sigma_2 = \sigma, \\ \sqrt{2}\log\frac{\tan\frac{a_1}{2}}{\tan\frac{a_2}{2}} & \text{otherwise.} \end{cases}$$

where $a_1 = \arcsin\frac{\sigma_1}{b_{12}}$, $a_2 = \arcsin\frac{\sigma_2}{b_{12}}$ and

$$b_{12} = \sigma_1^2 + \frac{(\mu_1 - \mu_2)^2 - 2(\sigma_2^2 - \sigma_1^2)}{8(\mu_1 - \mu_2)^2}.$$

For univariate normal distributions, Rao's distance amounts to computing the hyperbolic distance for $\mathbb{H}(\frac{1}{\sqrt{2}})$, see [34].

Statistical distances play a key role in tests of significance and classification [43]. Rao's distance is a metric since it is a Riemannian geodesic distance, and thus satisfies the triangle inequality. Rao's Riemannian geometric modeling of the population space is now commonly called the Fisher-Rao geometry [37]. One drawback of the Fisher-Rao geometry is the computer tractability of dealing with Riemannian geodesics. The following section concisely reviews the field of information geometry.

3. A Brief Overview of Information Geometry

Since the seminal work of Rao [6] in 1945, the interplay of differential geometry with statistics has further strengthened and developed into a new discipline called *information geometry* with a few dedicated monographs [5, 40, 30, 6, 46, 7]. It has been proved by Chentsov and published in his Russian monograph in 1972 (translated in English in 1982 by the AMS [18]) that the Fisher information matrix is the *only* invariant Riemannian metric for statistical manifolds (up to some scalar factor). Furthermore, Chentsov [18] proved that there exists a family of connections, termed the α-connections, that ensures statistical invariance.

3.1. Statistical Invariance and f-divergences.
A divergence is basically a smooth statistical distance that may not be symmetric nor satisfy the triangle inequality. We denote by $D(p : q)$ the divergence from distribution $p(x)$ to distribution $q(x)$, where the ":" notation emphasizes the fact that this dissimilarity measure may not be symmetric: $D(p : q) \neq D(q : p)$.

It has been proved that the only statistical invariant divergences [6, 42] are the Ali-Silvey-Csiszár f-divergences D_f [1, 21] that are defined for a functional convex generator f satisfying $f(1) = f'(1) = 0$ and $f''(1) = 1$ by:

$$D_f(p : q) = \int_{x \in \mathcal{X}} p(x) f\left(\frac{q(x)}{p(x)}\right) \, \mathrm{d}x.$$

Indeed, under an invertible mapping function (with $\dim \mathcal{X} = \dim \mathcal{Y} = d$):

$$\begin{aligned} m : \quad & \mathcal{X} \rightarrow \mathcal{Y} \\ & x \mapsto y = m(x) \end{aligned}$$

a probability density $p(x)$ is converted into another density $q(y)$ such that:

$$p(x)\mathrm{d}x = q(y)\mathrm{d}y, \qquad \mathrm{d}y = |M(x)|\mathrm{d}x,$$

26

where $|M(x)|$ denotes the determinant of the Jacobian matrix [6] of the transformation m (i.e., the partial derivatives):

$$M(x) = \begin{bmatrix} \frac{\partial y_1}{\partial x_1} & \cdots & \frac{\partial y_1}{\partial x_d} \\ \vdots & \ddots & \vdots \\ \frac{\partial y_d}{\partial x_1} & \cdots & \frac{\partial y_d}{\partial x_d} \end{bmatrix}.$$

It follows that

$$q(y) = q(m(x)) = p(x)|M(x)|^{-1}.$$

For any two densities p_1 and p_2, we have the f-divergence on the transformed densities q_1 and q_2 that can be rewritten mathematically as

$$\begin{aligned} D_f(q_1 : q_2) &= \int_{y \in \mathcal{Y}} q_1(y) f\left(\frac{q_2(y)}{q_1(y)}\right) dy, \\ &= \int_{x \in \mathcal{X}} p_1(x)|M(x)|^{-1} f\left(\frac{p_2(x)}{p_1(x)}\right) |M(x)| dx, \\ &= D_f(p_1 : p_2). \end{aligned}$$

Furthermore, the f-divergences are the only divergences satisfying the remarkable data-processing theorem [24] that characterizes the property of information monotonicity [4]. Consider discrete distributions on an alphabet $\mathcal{X}$ of d letters. For any partition $\mathcal{B} = \mathcal{X}_1 \cup ... \mathcal{X}_b$ of $\mathcal{X}$ that merge alphabet letters into $b \leq d$ bins, we have

$$0 \leq D_f(\bar{p}_1 : \bar{p}_2) \leq D_f(p_1 : p_2),$$

where $\bar{p}_1$ and $\bar{p}_2$ are the discrete distribution induced by the partition $\mathcal{B}$ on $\mathcal{X}$. That is, we loose discrimination power by coarse-graining the support of the distributions.

The most fundamental f-divergence is the Kullback-Leibler divergence [19] obtained for the generator $f(x) = x \log x$:

$$KL(p : q) = \int p(x) \log \frac{p(x)}{q(x)} dx.$$

The Kullback-Leibler divergence between two distributions $p(x)$ and $q(x)$ is equal to the cross-entropy $H^\times(p : q)$ minus the Shannon entropy $H(p)$:

$$\begin{aligned} KL(p : q) &= \int p(x) \log \frac{p(x)}{q(x)} dx, \\ &= H^\times(p : q) - H(p). \end{aligned}$$

27

with

$$H^{\times}(p:q) \;=\; \int -p(x)\log q(x)\mathrm{d}x,$$

$$H(p) \;=\; \int -p(x)\log p(x)\mathrm{d}x = H^{\times}(p:p).$$

The Kullback-Leibler divergence $\mathrm{KL}(\tilde{p}:p)$ [19] can be interpreted as the distance between the estimated distribution $\tilde{p}$ (from the samples) and the true hidden distribution p.

3.2. Information and Sufficiency.

In general, statistical invariance is characterized under Markov morphisms [38, 42] (also called sufficient stochastic kernels [42]) that generalizes the deterministic transformations $y = m(x)$. Loosely speaking, a geometric parametric statistical manifold $\mathcal{F} = \{p_\theta(x)|\theta \in \Theta\}$ equipped with a f-divergence must also provide invariance by:

Non-singular parameter reparameterization. That is, if we choose a different coordinate system, say $\theta' = f(\theta)$ for an invertible transformation f, it should not impact the intrinsic distance between the underlying distributions. For example, whether we parametrize the Gaussian manifold by $\theta = (\mu, \sigma)$ or by $\theta' = (\mu^3, \sigma^2)$, it should preserve the distance.

Sufficient statistic. When making statistical inference, we use statistics $T :$ $\mathbb{R}^d \to \Theta \subseteq \mathbb{R}^D$ (e.g., the mean statistic $T_n(X) = \frac{1}{n}\sum_{i=1}^n X_i$ is used for estimating the parameter μ of Gaussians). In statistics, the concept of *sufficiency* was introduced by Fisher [26]:

"... the statistic chosen should summarize the whole of the relevant information supplied by the sample. "

Mathematically, the fact that all information should be aggregated inside the sufficient statistic is written as

$$\Pr(x|t,\theta) = \Pr(x|t).$$

It is not surprising that all statistical information of a parametric distribution with D parameters can be recovered from a set of D statistics. For example, the univariate Gaussian with $d = \dim \mathcal{X} = 1$ and $D = \dim \Theta = 2$ (for parameters $\theta = (\mu, \sigma)$) is recovered from the mean and variance statistics. A sufficient statistic is a set of statistics that compress information without loss for statistical inference.

3.3. Sufficiency and Exponential Families. The distributions admitting finite sufficient statistics are called the exponential families [31, 14, 6], and have their probability density or mass functions canonically rewritten as

$$p_\theta(x) = \exp(\theta^\top t(x) - F(\theta) + k(x)),$$

where $k(x)$ is an auxiliary carrier measure, $t(x) : \mathbb{R}^d \to \mathbb{R}^D$ is the sufficient statistics, and $F : \mathbb{R}^D \to \mathbb{R}$ a strictly convex and differentiable function, called the cumulant function or the log normalizer since,

$$F(\theta) = \log \int_{x \in \mathcal{X}} \exp(\theta^\top t(x) + k(x)) \mathrm{d}x.$$

See [6] for canonical decompositions of usual distributions (Gaussian, multinomial, etc.). The space Θ for which the log-integrals converge is called the natural parameter space.

For example,

- Poisson distributions are univariate exponential distributions of order 1 (with $\mathcal{X} = \mathbb{N}^* = \{0, 1, 2, 3, ...\}$ and $\dim \Theta = 1$) with associated probability mass function:

$$\frac{\lambda^k e^{-\lambda}}{k!},$$

for $k \in \mathbb{N}^*$.

The canonical exponential family decomposition yields

  - $t(x) = x$: the sufficient statistic,
  - $\theta = \log \lambda$: the natural parameter,
  - $F(\theta) = \exp \theta$: the cumulant function,
  - $k(x) = -\log x!$: the carrier measure.

- Univariate Gaussian distributions are distributions of order 2 (with $\mathcal{X} = \mathbb{R}$, $\dim \mathcal{X} = 1$ and $\dim \Theta = 2$), characterized by two parameters $\theta = (\mu, \sigma)$ with associated density:

$$\frac{1}{\sigma\sqrt{2\pi}} e^{-\frac{1}{2}\left(\frac{x-\mu}{\sigma}\right)^2},$$

for $x \in \mathbb{R}$.

The canonical exponential family decomposition yields:

  - $t(x) = (x, x^2)$: the sufficient statistic,
  - $\theta = (\theta_1, \theta_2) = (\frac{\mu}{\sigma^2}, -\frac{1}{2\sigma^2})$: the natural parameters,

- $F(\theta) = -\frac{\theta_1^2}{4\theta_2} + \frac{1}{2}\log\left(-\frac{\pi}{\theta_2}\right)$: the cumulant function,
- $k(x) = 0$: the carrier measure.

Exponential families provide a generic framework in Statistics, and are universal density approximators [2]. That is, any distribution can be arbitrarily approximated closely by an exponential family. An exponential family is defined by the functions $t(\cdot)$ and $k(\cdot)$, and a member of it by a natural parameter θ. The cumulant function F is evaluated by the log-Laplace transform.

To illustrate the generic behavior of exponential families in Statistics [14], let us consider the maximum likelihood estimator for a distribution belonging to the exponential family. We have the MLE $\hat{\theta}$:

$$\hat{\theta} = (\nabla F)^{-1}\left(\sum_{i=1}^{n}\frac{1}{n}t(x_i)\right),$$

where $(\nabla F)^{-1}$ denotes the reciprocal gradient of F: $(\nabla F)^{-1} \circ \nabla F = \nabla F \circ (\nabla F)^{-1} = \mathrm{Id}$, the identity function on $\mathbb{R}^D$. The Fisher information matrix of an exponential family is

$$I(\theta) = \nabla^2 F(\theta) \succ 0,$$

the Hessian of the log-normalizer, always positive-definite since F is strictly convex.

3.4. Dual Bregman Divergences and α-Divergences.
The Kullback-Leibler divergence between two distributions belonging to the same exponential families can be expressed equivalently as a Bregman divergence on the swapped natural parameters defined for the cumulant function F of the exponential family:

$$\begin{aligned}
\mathrm{KL}(p_{F,\theta_1}(x) : p_{F,\theta_2}(x)) &= B_F(\theta_2 : \theta_1), \\
&= F(\theta_2) - F(\theta_1) - (\theta_2 - \theta_1)^{\top}\nabla F(\theta_1)
\end{aligned}$$

As mentioned earlier, the ":" notation emphasizes that the distance is not a metric: It does not satisfy the symmetry nor the triangle inequality in general. Divergence B_F is called a Bregman divergence [13], and is the canonical distances of dually flat spaces [6]. This Kullback-Leibler divergence on densities $\leftrightarrow$ divergence on parameters relies on the dual canonical parameterization of exponential families [14]. A random variable $X \sim p_{F,\theta}(x)$, whose distribution belongs to an exponential family, can be dually indexed by its expectation parameter η such that

$$\eta = E[t(X)] = \int_{x\in\mathcal{X}} xe^{\theta^{\top}t(x)-F(\theta)+k(x)}\mathrm{d}x = \nabla F(\theta).$$

For example, the η-parameterization of Poisson distribution is: $\eta = \nabla F(\theta) = e^{\theta} = \lambda = E[X]$ (since $t(x) = x$).

In fact, the Legendre-Fenchel convex duality is at the heart of information geometry: Any strictly convex and differentiable function F admits a dual convex conjugate F^* such that:

$$F^*(\eta) = \max_{\theta \in \Theta} \theta^{\top} \eta - F(\theta).$$

The maximum is attained for $\eta = \nabla F(\theta)$ and is unique since $F(\theta)$ is strictly convex ($\nabla^2 F(\theta) \succ 0$). It follows that $\theta = \nabla F^{-1}(\eta)$, where ∇F^{-1} denotes the functional inverse gradient. This implies that:

$$F^*(\eta) = \eta^{\top}(\nabla F)^{-1}(\eta) - F((\nabla F)^{-1}(\eta)).$$

The Legendre transformation is also called slope transformation since it maps $\theta \rightarrow \eta = \nabla F(\theta)$, where $\nabla F(\theta)$ is the gradient at θ, visualized as the slope of the support tangent plane of F at θ. The transformation is an involution for strictly convex and differentiable functions: $(F^*)^* = F$. It follows that gradient of convex conjugates are reciprocal to each other: $\nabla F^* = (\nabla F)^{-1}$. Legendre duality induces dual coordinate systems:

$$\eta = \nabla F(\theta),$$
$$\theta = \nabla F^*(\eta).$$

Furthermore, those dual coordinate systems are orthogonal to each other since,

$$\nabla^2 F(\theta) \nabla^2 F^*(\eta) = \text{Id},$$

the identity matrix.

The Bregman divergence can also be rewritten in a canonical mixed coordinate form C_F or in the θ- or η-coordinate systems as

$$\begin{aligned} B_F(\theta_2 : \theta_1) &= F(\theta_2) + F^*(\eta_1) - \theta_2^{\top} \eta_1 = C_F(\theta_2, \eta_1) = C_{F^*}(\eta_1, \theta_2), \\ &= B_{F^*}(\eta_1 : \eta_2). \end{aligned}$$

Another use of the Legendre duality is to interpret the log-density of an exponential family as a dual Bregman divergence [9]:

$$\log p_{F,t,k,\theta}(x) = -B_{F^*}(t(x) : \eta) + F^*(t(x)) + k(x),$$

with $\eta = \nabla F(\theta)$ and $\theta = \nabla F^*(\eta)$.

The Kullback-Leibler divergence (a f-divergence) is a particular divergence belonging to the 1-parameter family of divergences, called α-divergences (see [6], p. 57). The α-divergences are defined for $\alpha \neq \pm 1$ as

$$D_\alpha(p:q) = \frac{4}{1-\alpha^2}\left(1 - \int p(x)^{\frac{1-\alpha}{2}} q(x)^{\frac{1+\alpha}{2}} \mathrm{d}x\right).$$

It follows that $D_\alpha(q:p) = D_{-\alpha}(p:q)$, and in the limit case, we have:

$$D_{-1}(p:q) = \mathrm{KL}(p:q) = \int p(x) \log \frac{p(x)}{q(x)} \mathrm{d}x.$$

(Divergence D_1 is also called the reverse Kullback-Leibler divergence, and divergence D_0 the squared Hellinger distance.) In the sequel, we denote by D the divergence D_{-1} corresponding to the Kullback-Leibler divergence.

3.5. Exponential Geodesics and Mixture Geodesics.

Information geometry as further pioneered by Amari [6] considers dual affine geometries introduced by a pair of connections: the α-connection and $-\alpha$-connection instead of taking the Levi-Civita connection induced by the Fisher information Riemmanian metric of Rao. The ± 1-connections give rise to dually flat spaces [6] equipped with the Kullback-Leibler divergence [19]. The case of $\alpha = -1$ denotes the mixture family, and the exponential family is obtained for $\alpha = 1$. We omit technical details in this expository paper, but refer the reader to the monograph [6] for details.

For our purpose, let us say that the geodesics are defined not anymore as shortest path lengths (like in the metric case of the Fisher-Rao geometry) but rather as curves that ensures the parallel transport of vectors [6]. This defines the notion of "straightness" of lines. Riemannian geodesics satisfy both the straightness property and the minimum length requirements. Introducing dual connections, we do not have anymore distances interpreted as curve lengths, but the geodesics defined by the notion of straightness only.

In information geometry, we have dual geodesics that are expressed for the exponential family (induced by a convex function F) in the dual affine coordinate systems θ/η for $\alpha = \pm 1$ as:

$$\gamma_{12} \quad : \quad L(\theta_1, \theta_2) = \{\theta = (1-\lambda)\theta_1 + \lambda\theta_2 \mid \lambda \in [0,1]\},$$
$$\gamma_{12}^* \quad : \quad L^*(\eta_1, \eta_2) = \{\eta = (1-\lambda)\eta_1 + \lambda\eta_2 \mid \lambda \in [0,1]\}.$$

Furthermore, there is a Pythagorean theorem that allows one to define information-theoretic projections [6]. Consider three points p, q and r such that γ_{pq} is the θ-geodesic linking p to q, and γ_{qr}^* is the η-geodesic linking q

to r. The geodesics are orthogonal at the intersection point q if and only if the Pythagorean relation is satisfied:

$$D(p : r) = D(p : q) + D(q : r).$$

In fact, a more general triangle relation (extending the law of cosines) exists:

$$D(p : q) + D(q : r) - D(p : r) = (\theta(p) - \theta(q))^\top (\eta(r) - \eta(q)).$$

Note that the θ-geodesic γ_{pq} and η-geodesic γ_{qr}^* are orthogonal with respect to the inner product $G(q)$ defined at q (with $G(q) = I(q)$ being the Fisher information matrix at q). Two vectors u and v in the tangent place T_q at q are said to be orthogonal if and only if their inner product equals zero:

$$u \perp_q v \Leftrightarrow u^\top I(q)v = 0.$$

Observe that in any tangent plane T_x of the manifold, the inner product induces a squared Mahalanobis distance:

$$D_x(p, q) = (p - q)^\top I(x)(p - q).$$

Since $I(x) \succ 0$ is positive definite, we can apply Cholesky decomposition on the Fisher information matrix $I(x) = L(x)L^\top(x)$, where $L(x)$ is a lower triangular matrix with strictly positive diagonal entries.

By mapping the points p to $L(p)^\top$ in the tangent space T_p, the squared Mahalanobis amounts to computing the squared Euclidean distance $D_E(p, q) = \|p - q\|^2$ in the tangent planes:

$$
\begin{aligned}
D_x(p, q) &= (p - q)^\top I(x)(p - q), \\
&= (p - q)^\top L(x)L^\top(x)(p - q), \\
&= D_E(L^\top(x)p, L^\top(x)q).
\end{aligned}
$$

It follows that after applying the "Cholesky transformation" of objects into the tangent planes, we can solve geometric problems in tangent planes as one usually does in the Euclidean geometry.

Information geometry of dually flat spaces thus extend the traditional self-dual Euclidean geometry, obtained for the convex function $F(x) = \frac{1}{2}x^\top x$ (and corresponding to the statistical manifold of isotropic Gaussians).

4. Conclusion and Perspectives

Rao's paper [43] has been instrumental for the development of modern statistics. In this masterpiece, Rao introduced what is now commonly known as

33

the Cramér-Rao lower bound (CRLB) and the Fisher-Rao geometry. Both the contributions are related to the Fisher information, a concept due to Sir R. A. Fisher, the father of mathematical statistics [26] that introduced the concepts of consistency, efficiency and sufficiency of estimators. This paper is undoubtably recognized as the cornerstone for introducing differential geometric methods in Statistics. This seminal work has inspired many researchers and has evolved into the field of information geometry [6]. Geometry is originally the science of Earth measurements. But geometry is also the science of invariance as advocated by Felix Klein Erlang's program, the science of intrinsic measurement analysis. This expository paper has presented the two key contributions of C. R. Rao in his 1945 foundational paper, and briefly presented information geometry without the burden of differential geometry (e.g., vector fields, tensors, and connections). Information geometry has now ramified far beyond its initial statistical scope, and is further expanding prolifically in many different new horizons. To illustrate the versatility of information geometry, let us mention a few research areas:

- Fisher-Rao Riemannian geometry [37],

- Amari's dual connection information geometry [6],

- Infinite-dimensional exponential families and Orlicz spaces [16],

- Finsler information geometry [45],

- Optimal transport geometry [28],

- Symplectic geometry, Kähler manifolds and Siegel domains [10],

- Geometry of proper scoring rules [23],

- Quantum information geometry [29].

Geometry with its own specialized language, where words like distances, balls, geodesics, angles, orthogonal projections, etc., provides "thinking tools" (affordances) to manipulate non-trivial mathematical objects and notions. The richness of geometric concepts in information geometry helps one to reinterpret, extend or design novel algorithms and data-structures by enhancing creativity. For example, the traditional expectation-maximization (EM) algorithm [25] often used in Statistics has been reinterpreted and further extended using the framework of information-theoretic alternative projections [3]. In machine learning, the famous boosting technique that learns a strong classifier by combining linearly weak weighted classifiers has been revisited [39] under the framework of information geometry. Another striking example, is the study of the geometry of dependence and Gaussianity for Independent Component Analysis [15].

34

References

[1] Ali, S.M. and Silvey, S. D. (1966). A general class of coefficients of divergence of one distribution from another. *J. Roy. Statist. Soc. Series B* **28**, 131–142.

[2] Altun, Y., Smola, A. J. and Hofmann, T. (2004). Exponential families for conditional random fields. In *Uncertainty in Artificial Intelligence (UAI)*, pp. 2–9.

[3] Amari, S. (1995). Information geometry of the EM and em algorithms for neural networks. *Neural Networks* **8**, 1379–1408.

[4] Amari, S. (2009). Alpha-divergence is unique, belonging to both f-divergence and Bregman divergence classes. *IEEE Trans. Inf. Theor.* **55**, 4925–4931.

[5] Amari, S., Barndorff-Nielsen, O. E., Kass, R. E., Lauritzen, S.. L. and Rao, C. R. (1987). *Differential Geometry in Statistical Inference.* Lecture Notes-Monograph Series. Institute of Mathematical Statistics.

[6] Amari, S. and Nagaoka, H. (2000). *Methods of Information Geometry.* Oxford University Press.

[7] Arwini, K. and Dodson, C. T. J. (2008). *Information Geometry: Near Randomness and Near Independence.* Lecture Notes in Mathematics # 1953, Berlin: Springer.

[8] Atkinson, C. and Mitchell, A. F. S. (1981). Rao's distance measure. *Sankhyā Series A* **43**, 345–365.

[9] Banerjee, A., Merugu, S., Dhillon, I. S. and Ghosh, J. (2005). Clustering with Bregman divergences. *J. Machine Learning Res.* **6**, 1705–1749.

[10] Barbaresco, F. (2009). Interactions between symmetric cone and information geometries: Bruhat-Tits and Siegel spaces models for high resolution autoregressive Doppler imagery. In *Emerging Trends in Visual Computing* (F. Nielsen, Ed.) *Lecture Notes in Computer Science* # 5416, pp. 124–163. Berlin / Heidelberg: Springer.

[11] Bhatia, R. and Holbrook, J. (2006). Riemannian geometry and matrix geometric means. *Linear Algebra Appl.* **413**, 594–618.

[12] Bhattacharyya, A. (1943). On a measure of divergence between two statistical populations defined by their probability distributions. *Bull. Calcutta Math. Soc.* **35**, 99–110.

[13] Bregman, L. M. (1967). The relaxation method of finding the common point of convex sets and its application to the solution of problems in convex programming. *USSR Computational Mathematics and Mathematical Physics* **7**, 200–217.

[14] Brown, L. D. (1986). *Fundamentals of Statistical Exponential Families: with Applications in Statistical Decision Theory.* Institute of Mathematical Statistics, Hayworth, CA, USA.

[15] Cardoso, J. F. (2003). Dependence, correlation and Gaussianity in independent component analysis. *J. Machine Learning Res.* **4**, 1177–1203.

[16] Cena, A. and Pistone, G. (2007). Exponential statistical manifold. *Ann. Instt. Statist. Math.* **59**, 27–56.

[17] Champkin, J. (2011). C. R. Rao. *Significance* **8**, 175–178.

[18] Chentsov, N. N. (1982). *Statistical Decision Rules and Optimal Inferences.* Transactions of Mathematics Monograph, # 53 (Published in Russian in 1972).

[19] Cover, T. M. and Thomas, J. A. (1991). *Elements of Information Theory.* New York: Wiley.

[20] Cramér, H. (1946). *Mathematical Methods of Statistics.* NJ, USA: Princeton University Press.

[21] Csiszár, I. (1967). Information-type measures of difference of probability distributions and indirect observation. *Studia Scientia. Mathematica. Hungarica* **2**, 229–318.

[22] Darmois, G. (1945). Sur les limites de la dispersion de certaines estimations. *Rev. Internat. Stat. Instt.* **13**.

[23] Dawid, A. P. (2007). The geometry of proper scoring rules. *Ann. Instt. Statist. Math.* **59**, 77–93.

[24] del Carmen Pardo, M. C. and Vajda, I. (1997). About distances of discrete distributions satisfying the data processing theorem of information theory. *IEEE Trans. Inf. Theory* **43**, 1288–1293.

[25] Dempster, A. P., Laird, N. M. and Rubin, D. B. (1977). Maximum likelihood from incomplete data via the EM algorithm. *J. Roy. Statist. Soc. Series B* **39**, 1–38.

[26] Fisher, R. A. (1922). On the mathematical foundations of theoretical statistics. *Phil. Trans. Roy. Soc. London, A* **222**, 309–368.

[27] Fréchet, M. (1943). Sur l'extension de certaines évaluations statistiques au cas de petits échantillons. *Internat. Statist. Rev.* **11**, 182–205.

[28] Gangbo, W. and McCann, R. J. (1996). The geometry of optimal transportation. *Acta Math.* **177**, 113–161.

[29] Grasselli, M. R. and Streater, R. F. (2001). On the uniqueness of the Chentsov metric in quantum information geometry. *Infinite Dimens. Anal., Quantum Probab. and Related Topics* **4**, 173–181.

[30] Kass, R. E. and Vos, P. W. (1997). *Geometrical Foundations of Asymptotic Inference.* New York: Wiley.

[31] Koopman, B. O. (1936). On distributions admitting a sufficient statistic. *Trans. Amer. Math. Soc.* **39**, 399–409.

[32] Kotz, S. and Johnson, N. L. (Eds.) (1993). *Breakthroughs in Statistics: Foundations and Basic Theory,* Volume I. New York: Springer.

[33] Lehmann, E. L. and Casella, G. (1998). *Theory of Point Estimation* 2nd ed. New York: Springer.

[34] Lovric, M., Min-Oo, M. and Ruh, E. A. (2000). Multivariate normal distributions parametrized as a Riemannian symmetric space. *J. Multivariate Anal.* **74**, 36–48.

[35] Mahalanobis, P. C. (1936). On the generalized distance in statistics. *Proc. National Instt. Sci., India* **2**, 49–55.

[36] Mahalanobis, P. C. (1948). Historical note on the D^2-statistic. *Sankhyā* **9**, 237–240.

[37] Maybank, S., Ieng, S. and Benosman, R. (2011). A Fisher-Rao metric for paracatadioptric images of lines. *Internat. J. Computer Vision*, 1–19.

[38] Morozova, E. A. and Chentsov, N. N. (1991). Markov invariant geometry on manifolds of states. *J. Math. Sci.* **56**, 2648–2669.

[39] Murata, N., Takenouchi, T., Kanamori, T. and Eguchi, S. (2004). Information geometry of U-boost and Bregman divergence. *Neural Comput.* **16**, 1437–1481.

[40] Murray, M. K. and Rice, J. W. (1993). *Differential Geometry and Statistics*. Chapman and Hall/CRC.

[41] Peter, A. and Rangarajan, A. (2006). A new closed-form information metric for shape analysis. In *Medical Image Computing and Computer Assisted Intervention (MICCAI)* Volume 1, pp. 249–256.

[42] Qiao, Y. and Minematsu, N. (2010). A study on invariance of f-divergence and its application to speech recognition. *Trans. Signal Process.* **58**, 3884–3890.

[43] Rao, C. R. (1945). Information and the accuracy attainable in the estimation of statistical parameters. *Bull. Calcutta Math. Soc.* **37**, 81–89.

[44] Rao, C. R. (2010). Quadratic entropy and analysis of diversity. *Sankhyā, Series A*, **72**, 70–80.

[45] Shen, Z. (2006). Riemann-Finsler geometry with applications to information geometry. *Chinese Annals of Mathematics* **27B**, 73–94.

[46] Shima, H. (2007). *The Geometry of Hessian Structures*. Singapore: World Scientific.

[47] Watanabe, S. (2009). *Algebraic Geometry and Statistical Learning Theory*. Cambridge: Cambridge University Press.

Frobenius Splittings

Wilberd van der Kallen[*]

1. Introduction

Frobenius splittings were introduced by V. B. Mehta and A. Ramanathan in [6] and refined further by S. Ramanan and Ramanathan in [9]. Frobenius splittings have proven to be a amazingly effective when they apply. Proofs involving Frobenius splittings tend to be very efficient. Other methods usually require a much more detailed knowledge of the object under study. For instance, while showing that the intersection of one union of Schubert varieties with another union of Schubert varieties is reduced, one does not need to know where that intersection is situated, let alone what it looks like exactly.

Before getting to serious applications we slowly introduce the main concepts.

2. Frobenius Splittings for Algebras

Fix a prime $p > 0$. Let A be commutative ring of characteristic p. So A contains the field $\mathbb{F}_p$ with p elements. The Frobenius homomorphism $\phi : A \to A$ is the ring map sending a to a^p. The same notation ϕ will be used for the Frobenius homomorphism on other $\mathbb{F}_p$-algebras. Let $\mathbb{F}$ be a perfect field of characteristic p. So the Frobenius map $\phi : \mathbb{F} \to \mathbb{F}$ is a field automorphism. The field $\mathbb{F}$ will serve as our base field.

Pull back   If M is an A-module, then $\phi^* M$ denotes the A-module obtained by *base change* along ϕ. That is, as an abelian group $\phi^* M$ equals M, but there is a different module structure, given as follows. Let us use $\Diamond$ to denote the new module structure. One puts

$$a \Diamond m := a^p m \quad \text{for } a \in A, \ m \in \phi^* M.$$

If we interpret $\phi : A \to A$ as a map $\phi : A \to \phi^* A$, then ϕ is A-linear:

$$\phi(a + b) = \phi(a) + \phi(b), \qquad \phi(ab) = a \Diamond \phi(b).$$

Splitting   Define a *Frobenius splitting* on A to be an A-linear map $\sigma : \phi^* A \to A$ with $\sigma \circ \phi = \mathrm{id}$. In other words, σ is a left inverse of ϕ, whence the name

---

[*]Mathematisch Instituut, Universiteit Utrecht, 3584 CD UTRECHT, The Netherlands

Frobenius splitting. We often just say *splitting*. When A has a splitting σ we call A or $\text{Spec}(A)$ *split* by σ.

A Frobenius splitting σ of A is just a set map σ from A to itself, satisfying

1. $\sigma(a+b) = \sigma(a) + \sigma(b)$, for $a, b \in A$,

2. $\sigma(a \Diamond b) = a\sigma(b)$, for $a, b \in A$,

3. $\sigma(1) = 1$.

Notice that these three properties do imply $\sigma(\phi(a)) = a$, because $\sigma(\phi(a)) = \sigma(a \Diamond 1) = a\sigma(1) = a$.

Call a map $\sigma : A \to A$ a *twisted* linear endomorphism if it satisfies (1) and (2). Write $\text{End}_\phi(A)$ for the abelian group of twisted linear endomorphism of A. We make it into an A-module by putting $(a * \sigma)(b) = \sigma(ab)$ for $a \in A$, $\sigma \in \text{End}_\phi(A)$, $b \in A$. So the module structure on $\text{End}_\phi(A)$ is given by premultiplication. Postmultiplication as in $(a\sigma)(b) = a\sigma(b)$ describes the A-module structure on $\phi^* \text{End}_\phi(A)$.

Here is the first result. Recall that a ring is *reduced* if it has no nonzero *nilpotent* elements [2, p. 33].

Lemma 2.1. *If A has a Frobenius splitting, then A is reduced.*

Proof If not, there is an $a \in A$, $a \neq 0$ with $a^2 = 0$. But then $a = \sigma(\phi(a)) = \sigma(a^p) = \sigma(0) = 0$. Contradiction. $\qquad\square$

We can see this lemma as a first indication that possessing a Frobenius splitting is something special. After all, not all A are reduced.

Polynomial rings We wish to understand $\text{End}_\phi(A)$ when A is a polynomial ring $\mathbb{F}[x_1, \ldots, x_n]$ over our perfect field $\mathbb{F}$. Let us start with the one variable case $A = \mathbb{F}[x]$. The A-module $\phi^* A$ has a basis $1, x, \ldots, x^{p-1}$, so $\sigma \in \text{End}_\phi(A)$ is determined by the $\sigma(x^i)$ with $i = 0, \ldots, p-1$. Define $\sigma_0 \in \text{End}_\phi(A)$ by stipulating that $\sigma_0(x^{p-1}) = 1$, $\sigma_0(x^i) = 0$ for $0 \leq i < p-1$.

Lemma 2.2. $\text{End}_\phi(\mathbb{F}[x])$ *is free with basis σ_0.*

Proof Let $\sigma \in \text{End}_\phi(\mathbb{F}[x])$. Put $f_i = \sigma(x^i)$ for $0 \leq i \leq p-1$. We claim that $\sigma = \sum_{i=0}^{p-1}(f_i \Diamond x^{p-1-i}) * \sigma_0$. Indeed, for $0 \leq j \leq p-1$ one gets

$$\left(\sum_{i=0}^{p-1}(f_i \Diamond x^{p-1-i}) * \sigma_0 \right)(x^j) \quad =$$

39

$$\sum_{i=0}^{p-1} \sigma_0(f_i \lozenge x^{p-1-i} x^j) =$$

$$\sum_{i=0}^{p-1} f_i \sigma_0(x^{p-1-i} x^j) = f_j \qquad \square$$

Tensor products Let A, B be $\mathbb{F}$-algebras. As $\mathbb{F}$ is perfect, there is a natural map from $\mathrm{End}_\phi(A) \otimes_\mathbb{F} \mathrm{End}_\phi(B)$ to $\mathrm{End}_\phi(A \otimes_\mathbb{F} B)$. For $\sigma \in \mathrm{End}_\phi(A)$, $\tau \in \mathrm{End}_\phi(B)$, $a \in A$, $b \in B$, we put $(\sigma \otimes \tau)(a \otimes b) = \sigma(a) \otimes \tau(b)$. This defines a twisted endomorphism $\sigma \otimes \tau$ of $A \otimes_\mathbb{F} B$.

Exercise 2.3. *If $A = \mathbb{F}[x_1, \ldots, x_n]$ then $\mathrm{End}_\phi(A)$ is free with basis σ_0, where $\sigma_0(x_1^{p-1} \cdots x_n^{p-1}) = 1$, while $\sigma_0(x_1^{m_1} \cdots x_n^{m_n}) = 0$ if at least one $m_i + 1$ is not divisible by p.*

Exercise 2.4. *The algebra $A = \mathbb{F}[x_1, \ldots, x_n]$ is graded with each x_i having degree one. The element σ_0 of the previous exercise sends homogeneous polynomials to homogeneous polynomials. If $f \in A$ is homogeneous, then $f * \sigma_0$ also sends homogeneous polynomials to homogeneous polynomials. In particular, if $(f * \sigma_0)(1)$ has constant term 1 and f is homogeneous, then $f * \sigma_0$ is a splitting.*

Lemma 2.5. *The following are equivalent*

- *$f * \sigma_0$ is a splitting,*

- *The coefficient of $x_1^{p-1} \cdots x_n^{p-1}$ in f is one, and the other monomials $x_1^{m_1} \cdots x_n^{m_n}$ with nonzero coefficient in f have at least one $m_i + 1$ not divisible by p.* $\square$

Remark 2.6. *The coefficient of $x_1^{p-1} \cdots x_n^{p-1}$ in f is the value of $(f * \sigma_0)(1)$ at the origin.*

Compatible ideal Let $\sigma \in \mathrm{End}_\phi(A)$ and let I be an ideal of A. We say that σ is *compatible* with I if $\sigma(I) \subset I$.

Write $\mathrm{End}_\phi(A, I)$ for $\{ \sigma \in \mathrm{End}_\phi(A) \mid \sigma(I) \subset I \}$. Clearly, if σ is compatible with I it induces a map $\bar\sigma : A/I \to A/I$ that also satisfies (1) and (2). So we get a map $\mathrm{End}_\phi(A, I) \to \mathrm{End}_\phi(A/I)$. It sends splittings to splittings.

Lemma 2.7. *If A has a Frobenius splitting compatible with I, then I is a radical ideal.*

Proof Indeed, A/I is reduced by Lemma 2.1. $\square$

Localization   If S is a multiplicatively closed subset of A, not containing zero, consider the *localization* $S^{-1}A$ of A [2, 2.1]. Recall that an element of $S^{-1}A$ may be written in more than one way as a fraction a/b. There is a natural localization map $\text{End}_\phi(A) \to \text{End}_\phi(S^{-1}A)$, say $\sigma \mapsto \sigma_S$, where $\sigma_S(a/b) = \sigma(ab^{p-1})/b$ for $a \in A$, $b \in S$. Check that σ_S is well defined. The localization map sends splittings to splittings. If S contains no zero divisors, then A is a subring of $S^{-1}A$ and σ is the restriction of σ_S to A.

Completion   If the ideal I is finitely generated, then one checks that any $\sigma \in \text{End}_\phi(A)$ is continuous for the I-adic topology, also known as the *Krull topology* [2, 7.5]. If $\hat{A}$ denotes the I-adic completion we get a map $\text{End}_\phi(A) \to \text{End}_\phi(\hat{A})$. It sends splittings to splittings.

Lemma 2.8. *Let $f \in A$ be a non zero divisor. Then*

$$\text{End}_\phi(A, (f)) = f^{p-1} * \text{End}_\phi(A).$$

Proof   On the one hand, if $\sigma \in \text{End}_\phi(A)$, then $(f^{p-1} * \sigma)(fa) = \sigma(f * a) = f\sigma(a)$ for $a \in A$, so that $f^{p-1} * \sigma \in \text{End}_\phi(A, (f))$. On the other hand, if $\sigma \in \text{End}_\phi(A, (f))$ define $\tau : A \to A$ by $\tau(a) = \sigma(fa)/f$. One checks that $\tau \in \text{End}_\phi(A)$ and that $f^{p-1} * \tau = \sigma$.   □

Example 2.9. *The cross is split.*

   Let $A = \mathbb{F}[x, y]$ be the polynomial ring in two variables. The splitting $\sigma = (xy)^{p-1} * \sigma_0$ is compatible with the ideal (xy). Indeed, $\sigma(xyf) = \sigma_0(x^p y^p f) = xy\sigma_0(f)$ for $f \in A$. So we have found a splitting on the coordinate ring $\mathbb{F}[x, y]/(xy)$ of the union of the x-axis and the y-axis. This coordinate ring is not normal [2, 4.2]. The normalization [2, 4.2] is $\mathbb{F}[x] \times \mathbb{F}[y]$, and the map from the spectrum [2, p. 54] of $\mathbb{F}[x] \times \mathbb{F}[y]$ to the spectrum of $\mathbb{F}[x, y]/(xy)$ pinches together two points. So a Frobenius splitting does not rule out such behaviour. However, it does rule out pinching together two infinitely near points as displayed in the next example.

Example 2.10. *The cusp is not split.*

Consider the subring $A = \mathbb{F}[t^2, t^3]$ of the polynomial ring $\mathbb{F}[t]$. It is the coordinate ring of a cusp. The polynomial ring $\mathbb{F}[t]$ is the normalization of A. The ideal $\mathfrak{c}$ generated by t^2 and t^3 in $\mathbb{F}[t]$ is the conductor ideal [2, Exercise 11.16]. It is a common ideal in A and in $\mathbb{F}[t]$. The ring $A/\mathfrak{c}$ is nonreduced. We already know that Frobenius splittings have little tolerance for nilpotents. So let us show that A cannot have a splitting. Suppose it did have a splitting σ. Take for S the set of nonzero elements of A. The splitting σ_S on the field of fractions $\mathbb{F}(t)$ must send t^p to t. But it also should send A to A. Now t^p is in A, but t is not. Contradiction.

Example 2.11. The node is split.

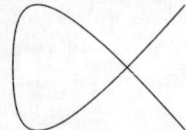

Let our prime p be unequal to two. Consider the ring $A = \mathbb{F}[x, y]/(y^2 - x^3 - x^2)$, the coordinate ring of an ordinary node. (In characteristic two the equation $y^2 = x^3 + x^2$ would define a cusp.)

One may check by direct computation that $(y^2 - x^3 - x^2)^{p-1} * \sigma_0$ is a splitting. So the ring A is Frobenius split compatibly with the ideal $(y^2 - x^3 - x^2)$. For many purposes it is good enough to know just the existence of a splitting in $\mathrm{End}_\phi(\mathbb{F}[x, y], (y^2 - x^3 - x^2))$. So then one would like to know that $1 \in \mathbb{F}[x, y]$ is being hit by the map from $\phi^* \mathrm{End}_\phi(\mathbb{F}[x, y], (y^2 - x^3 - x^2))$ to $\mathbb{F}[x, y]$ which sends σ to $\sigma(1)$. This is a linear algebra problem over a ring, so one has the usual tools of localization and completion at one's disposal. But after localization and completion we see no difference between the node and the cross [2, Second Example in 7.2]. So this explains why the ideal of the node in the plane is compatibly split.

Example 2.12. No splitting when there is higher order contact.

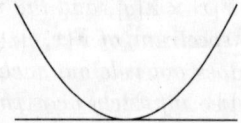

We now look at the ideal $I = (y(y - x^2)) \subset \mathbb{F}[x, y]$ of the union in the plane of the x-axis and the parabola $y = x^2$. We claim this ideal is not compatibly split, the reason being the higher order contact at the intersection of the parabola with the x-axis. Suppose $\sigma \in \mathrm{End}_\phi(\mathbb{F}[x, y], I)$ were a splitting. First let S consist of the powers of $y - x^2$. Then $\sigma_S \in \mathrm{End}_\phi(S^{-1}\mathbb{F}[x, y], S^{-1}I)$ maps $S^{-1}I$ to itself and also $\mathbb{F}[x, y]$ to itself.

We claim the intersection of $\mathbb{F}[x,y]$ with $S^{-1}I$ is the ideal (y) of $\mathbb{F}[x,y]$. Indeed, a polynomial function on the plane that vanishes on an open dense subset of the x-axis vanishes on the whole x-axis. Now take as open dense subset the intersection with the complement of the parabola. So σ is compatible with the ideal (y) of $\mathbb{F}[x,y]$. Similarly, by inverting y instead of $y - x^2$ we learn that σ is compatible with the ideal $(y - x^2)$ in $\mathbb{F}[x,y]$ of the other component. But then it must be compatible with the ideal $J = (y) + (y - x^2)$, the ideal of the scheme theoretic intersection of the two components. However, because of the higher order contact, this scheme theoretic intersection is not reduced: $\mathbb{F}[x,y]/J \cong \mathbb{F}[x]/(x^2)$ contains a nontrivial nilpotent.

Discussion What the last example shows us is that a Frobenius splitting allows to extrapolate from generic information, on dense subsets of components, to information about a special locus. As V. B. Mehta explained it to me, a splitting seems to make bad behaviour at special points spread out to a neighborhood of the bad point, which then makes it detectable generically. So behaviour that is not allowed generically gets forbidden everywhere. In local coordinates one may think of the Frobenius map $t \mapsto t^p$ as a concentration flow, so that the splitting becomes a diffusion flow.

3. Frobenius Splittings for Varieties

For simplicity we take our base field $\mathbb{F}$ algebraically closed, still of characteristic p, $p > 0$. We will consider *varieties* over $\mathbb{F}$, or more generally *schemes* over $\mathbb{F}$ [3]. Unlike [1] or [3] we do not require varieties to be smooth or connected. If X is a variety over $\mathbb{F}$ we do require that X is reduced and that the corresponding morphism $X \to \mathrm{Spec}(\mathbb{F})$ is of *finite type* [3, p. 84].

Frobenius map for varieties So let X be a variety or scheme over $\mathbb{F}$ with structure sheaf $\mathcal{O}_X$. The absolute Frobenius map $F : X \to X$ is the morphism of ringed spaces which is the identity on the underlying topological space and for any open subset U raises a section $f \in \Gamma(U, \mathcal{O}_X)$ to its p-th power. So we use F as notation for a Frobenius map of schemes and ϕ for a Frobenius map of algebras. Note that F is a map and that $\mathbb{F}$ is a field. Just like ϕ has been used for any Frobenius map of algebras, F will be used for any Frobenius map of schemes.

Example 3.1. *Let A be an $\mathbb{F}$-algebra and ϕ its Frobenius endomorphism. The corresponding morphism $\mathrm{Spec}(A) \to \mathrm{Spec}(A)$ is the absolute Frobenius map F.*

Say X is a variety over $\mathbb{F}$. If one wants to view the morphism of ringed spaces $F : X \to X$ as a morphism of varieties over $\mathbb{F}$, then one may exploit the following commutative diagram of schemes in which the vertical maps encode the $\mathbb{F}$-structure on X.

$$
\begin{array}{ccc}
X & \xrightarrow{F} & X \\
\downarrow & \searrow & \downarrow \\
\end{array}
$$
$$
\operatorname{Spec}(\mathbb{F}) \xrightarrow{F} \operatorname{Spec}(\mathbb{F})
$$

It suggests to view the source of $F : X \to X$ in a different way as a variety over $\mathbb{F}$, namely by using the diagonal map instead of the vertical one. The new variety structure obtained this way we denote $X^{(-1)}$. Then $F : X^{(-1)} \to X$ is a map of varieties over $\mathbb{F}$.

Splitting of a variety Let X be a variety or scheme over $\mathbb{F}$. Consider the sheaf map $\mathcal{O}_X \to F_*\mathcal{O}_X$. Over any open subset U it is described by the Frobenius map ϕ on the algebra $\Gamma(U, \mathcal{O}_X)$. Following Mehta and Ramanathan we define a *Frobenius splitting* σ on X to be a morphism of $\mathcal{O}_X$-modules $F_*\mathcal{O}_X \to \mathcal{O}_X$ that splits the map $\mathcal{O}_X \to F_*\mathcal{O}_X$. So the composite $\mathcal{O}_X \to F_*\mathcal{O}_X \xrightarrow{\sigma} \mathcal{O}_X$ must be the identity. A scheme with a Frobenius splitting is called *Frobenius split* or just *split*. Notice that the $\mathcal{O}_X$-module structure on $F_*\mathcal{O}_X$ is such that $\Gamma(U, F_*\mathcal{O}_X)$ as a $\Gamma(U, \mathcal{O}_X)$-module is the pull back module $\phi^*\Gamma(U, \mathcal{O}_X)$ in the notation of Section 2. So a Frobenius splitting σ on X is a sheaf map that gives a Frobenius splitting of each algebra $\Gamma(U, \mathcal{O}_X)$. Any open subset of a Frobenius split scheme is Frobenius split.

If A is an $\mathbb{F}$-algebra, then A is Frobenius split if and only if $\operatorname{Spec}(A)$ is Frobenius split. Thus Lemma 2.1 implies

Lemma 3.2. *A Frobenius split scheme is reduced.* □

Exercise 3.3. *The splitting $x^{p-1} * \sigma_0$ of $\mathbb{F}[x]$ obtained from Lemma 2.5 gives a splitting of $\mathbb{A}^1 = \operatorname{Spec}(\mathbb{F}[x])$ that extends to a splitting of $\mathbb{P}^1$.*

We write $\mathcal{E}nd_F(X)$ for the sheaf of abelian groups $\mathcal{H}om(F_*\mathcal{O}_X, \mathcal{O}_X)$ associated to the presheaf $U \mapsto \operatorname{End}_\phi(\Gamma(U, \mathcal{O}_X))$. We make it into an $\mathcal{O}_X$-module using the $\Gamma(U, \mathcal{O}_X)$-module structure $*$ on $\operatorname{End}_\phi(\Gamma(U, \mathcal{O}_X))$ from Section 2. (Recall that we view $\sigma \in \operatorname{End}_\phi(\Gamma(U, \mathcal{O}_X))$ as a map $\sigma : \Gamma(U, \mathcal{O}_X) \to \Gamma(U, \mathcal{O}_X)$ and put $(a * \sigma)(b) = \sigma(ab)$.) A splitting is a global section σ of $\mathcal{E}nd_F(X)$ with $\sigma(1) = 1$. If $\sigma(1)$ is some nonzero constant in $\mathbb{F}$, then we say that σ *spans a splitting*. The splitting it spans is of the form $\alpha * \sigma$ with $\alpha \in \mathbb{F}$. Put $\operatorname{End}_F(X) = \operatorname{Hom}(F_*\mathcal{O}_X, \mathcal{O}_X) = \Gamma(X, \mathcal{E}nd_F(X))$ and refer to its elements as *twisted endomorphisms* of $\mathcal{O}_X$.

Example 3.4. *If $A = \mathbb{F}[x_1, \ldots, x_n]$, then $\operatorname{Spec}(A) = \mathbb{A}^n$ is Frobenius split. The sheaf $\mathcal{E}nd_F(X)$ is a trivial line bundle with nowhere vanishing global section*

given by the element σ_0 from Exercise 2.3. Locally any smooth n-dimensional variety looks like $\mathbb{A}^n$ up to completion, so $\mathcal{E}nd_F(X)$ must be a line bundle on any smooth variety X. But there is no reason that it should be a trivial line bundle and in fact it is usually not. We already see a problem with $\mathbb{A}^n$ itself. If $\alpha_1, \ldots, \alpha_n \in \mathbb{F}$ are nonzero and $y_i = \alpha_i x_i$, then $\mathbb{F}[x_1, \ldots, x_n] = \mathbb{F}[y_1, \ldots, y_n]$. If σ_0^x denotes the generator of $\mathrm{End}_\phi(A)$ constructed with the x_i and σ_0^y is the one constructed with the y_i, then $\sigma_0^x = (\alpha_1 \cdots \alpha_n)^{p-1} * \sigma_0^y$. That does not look like the transformation behaviour for structure sheafs.

Theorem 3.5 (Mehta-Ramanathan). *If X is smooth of dimension n, then $\mathcal{E}nd_F(X)$ is isomorphic as a line bundle to ω_X^{1-p}, where ω_X is the canonical line bundle of n-forms on X and ω_X^{1-p} means $\mathcal{H}om(\omega_X^p, \omega_X)$.*

Remark 3.6. *So if $\Gamma(X, \omega_X^{1-p})$ vanishes, then X is certainly not split. Recall that a smooth complete variety is called a Fano variety if ω_X^{-1} is ample. While they are 'quite rare' in algebraic geometry, they are very common in representation theory of reductive algebraic groups. Indeed, that is where many applications of Frobenius splittings are to be found.*

The theorem is easy to prove with local duality theory, but we prefer to use the Cartier operator to construct a natural isomorphism between the line bundles $\mathcal{E}nd_F(X)$ and ω_X^{1-p}. That will open the way to some explicit computations in local coordinates. This is important for checking that a given global section of $\mathcal{E}nd_F(X)$ is a splitting. So let us digress and recall how the Cartier operator works.

4. Cartier Operator

Let X be a variety of dimension n over $\mathbb{F}$. We consider the DeRham complex

$$0 \to \mathcal{O}_X \to \Omega_X^1 \to \cdots \to \Omega_X^n \to 0$$

with as differential d the usual exterior differentiation. Because this differential is not $\mathcal{O}_X$-linear, we twist the $\mathcal{O}_X$-module structure on Ω_X^i by putting $f \lozenge \omega = f^p \omega$ for a section $f \in \Gamma(U, \mathcal{O}_X)$ and a differential i-form $\omega \in \Gamma(U, \Omega_X^i)$. With this twisted module structure the DeRham complex is a complex of coherent $\mathcal{O}_X$-modules, and the exterior algebra $\Omega_X^* = \bigoplus_{i=0}^n \Omega_X^i$ is a differential graded $\mathcal{O}_X$-algebra. We denote its cohomology sheafs $\mathcal{H}_{\mathrm{dR}}^i$. They are $\mathcal{O}_X$-modules by means of the twisted action. If U is an affine open subset, then $\Gamma(U, \mathcal{H}_{\mathrm{dR}}^i)$ consists of all closed differential i-forms on U modulo the exact ones. Now consider the map $\gamma : f \mapsto$ class of $f^{p-1} df$ from $\mathcal{O}_X$ to $\mathcal{H}_{\mathrm{dR}}^1$.

Lemma 4.1. *γ is a derivation and thus induces an $\mathcal{O}_X$-algebra homomorphism $c : \Omega_X^* \to \mathcal{H}_{\mathrm{dR}}^*$.*

45

Remark 4.2. *Note that one should put the ordinary $\mathcal{O}_X$-module structure on Ω_X^* here, not the twisted one that is used for $\mathcal{H}_{dR}^*$.*

Proof of Lemma 4.1 With

$$\Phi(X,Y) = ((X+Y)^p - X^p - Y^p)/p \in \mathbb{Z}[X,Y]$$

we get

$$
\begin{aligned}
(f+g)^{p-1}d(f+g) &= f^{p-1}df + g^{p-1}dg + d\Phi(f,g) \\
(fg)^{p-1}d(fg) &= g\Diamond f^{p-1}df + f\Diamond g^{p-1}dg,
\end{aligned}
$$

where the first equality is a consequence of the fact that

$$p(X+Y)^{p-1}d(X+Y) = pX^{p-1}dX + pY^{p-1}dY + pd\Phi(X,Y)$$

in the torsion free $\mathbb{Z}$-module $\Omega^1_{\mathbb{Z}[X,Y]}$. $\square$

Proposition 4.3. *If X is smooth, the homomorphism c is bijective.*

Cartier operator The inverse map $C : \mathcal{H}_{dR}^* \to \Omega_X^*$ is called the *Cartier operator* (cf. [8]).

Proof of Proposition 4.3 To check that a map of coherent sheafs is an isomorphism it suffices to check that one gets an isomorphism after passing to the completion at an arbitrary closed point. But then we are simply dealing with the DeRham complex for a power series ring in n variables over k and everything can be made very explicit (exercise). $\square$

Remark 4.4. *Here are some formulas satisfied by the Cartier operator, in informal notation. In view of these formulas the connection with Frobenius splittings is not surprising.*

- $C(f^p\tau) = fC(\tau)$

- $C(d\tau) = 0$

- $C(\mathrm{dlog}\, f) = \mathrm{dlog}\, f$, *where* $\mathrm{dlog}\, f$ *stands for* $(1/f)df$ *if* f *is invertible (or after f has been inverted).*

- $C(\xi \wedge \tau) = C(\xi) \wedge C(\tau)$

Here f is a function and ξ, τ are forms.

Proposition 4.5. *If X is smooth, we have a natural isomorphism of $\mathcal{O}_X$-modules*

$$\mathcal{E}nd_F(X) \cong \omega_X^{1-p} = \mathcal{H}om(\omega_X^p, \omega_X),$$

where ω_X is the canonical line bundle Ω_X^n. If τ is a local generator of ω_X, f a local section of $\mathcal{O}_X$, ψ a local homomorphism $\omega_X^p \to \omega_X$, then the local section σ of $\mathcal{E}nd_F(X)$ corresponding to ψ is defined by $\sigma(f)\tau = C(\text{class of } \psi(f\tau^{\otimes p}))$.

Proof One checks that $C(\text{class of } \psi(f\tau^{\otimes p}))/\tau$ does not depend on the choice of τ, so that σ depends only on ψ. To see that the map $\psi \mapsto \sigma$ defines an isomorphism of line bundles we may argue as in the previous proof. □

Remark 4.6. *If X is smooth of dimension zero, then $\omega_X = \mathcal{O}_X$ and Proposition 4.5 describes the isomorphism $\mathrm{End}_\phi(\mathbb{F}) \cong \mathbb{F}$.*

Example 4.7. *We try the Proposition out for $X = \mathbb{A}^n = \mathrm{Spec}(\mathbb{F}[x_1, \ldots, x_n])$. An obvious generator of ω_X is $\tau = dx_1 \wedge \cdots \wedge dx_n$. As local section of $\mathcal{O}_X$ at some point we take a global function f that does not vanish at the point. The most obvious generator of $\mathrm{Hom}(\omega_X^p, \omega_X)$ sends $\tau^{\otimes p}$ to τ. One may write it as τ^{1-p}. We claim it corresponds with our old friend σ_0 from Exercise 2.3. We must check that $\sigma_0(f)\tau = C(f\tau)$, in simplified notation. It suffices to consider the case where f is a monomial $x_1^{m_1} \cdots x_n^{m_n}$. If $m_n + 1$ not divisible by p, then $f\tau = dx_1 \wedge \cdots \wedge dx_{n-1} \wedge d(x_n f)/(m_n + 1)$ is a boundary so that $C(f\tau)$ vanishes. Similarly $C(f\tau)$ vanishes if some other $m_i + 1$ is not divisible by p. So far the results are consistent with the definition of σ_0. We still need to inspect the case where $x_1 \cdots x_n f$ is a p-th a power, say g^p. Then $C(f\tau) = C(g^p \operatorname{dlog} x_1 \wedge \cdots \wedge \operatorname{dlog} x_n) = g \operatorname{dlog} x_1 \wedge \cdots \wedge \operatorname{dlog} x_n = (x_1 \cdots x_n)^{-1} g\tau$. Indeed, $\sigma_0(f)$ equals $(x_1 \cdots x_n)^{-1} g$.*

5. Sheaf Cohomology

Let X be a variety over $\mathbb{F}$ and let $\mathcal{L}$ be a line bundle on X. Then $F^*\mathcal{L}$ is isomorphic to $\mathcal{L}^p$. On an affine open subset U, say $U = \mathrm{Spec}(A)$, the isomorphism sends $a \otimes s \in A \otimes_\phi \Gamma(U, \mathcal{L}) = \Gamma(U, F^*\mathcal{L})$ to $a(s^{\otimes p}) \in \Gamma(U, \mathcal{L}^p)$.

If $\mathcal{M}$ is a sheaf of abelian groups and $i \geq 0$ then $H^i(X, F_*\mathcal{M}) = H^i(X, \mathcal{M})$ as abelian groups, because F is the identity on the underlying topological space.

Now suppose that σ splits X. Then $\mathcal{L} \otimes_{\mathcal{O}_X} \mathcal{O}_X \to \mathcal{L} \otimes_{\mathcal{O}_X} F_*\mathcal{O}_X$ is split injective, so $H^i(X, \mathcal{L} \otimes_{\mathcal{O}_X} \mathcal{O}_X) \to H^i(X, \mathcal{L} \otimes_{\mathcal{O}_X} F_*\mathcal{O}_X)$ is split injective for each i. By the projection formula $\mathcal{L} \otimes_{\mathcal{O}_X} F_*\mathcal{O}_X$ equals $F_*(F^*\mathcal{L} \otimes_{\mathcal{O}_X} \mathcal{O}_X) = F_*\mathcal{L}^p$. We get split injective maps $H^i(X, \mathcal{L}) \to H^i(X, \mathcal{L}^p)$. By iteration we get split injective maps $H^i(X, \mathcal{L}) \to H^i(X, \mathcal{L}^{p^r})$ for $r \geq 1$.

Proposition 5.1 ([6, Proposition 1]). *Let X be a projective variety which is Frobenius split. Let $\mathcal{L}$ be a line bundle so that for some i, $H^i(X, \mathcal{L}^m) = 0$ for all large m (e.g. $i > 0$, $\mathcal{L}$ ample). Then $H^i(X, \mathcal{L}) = 0$.* $\square$

Proposition 5.2 (Kodaira's vanishing theorem [6, Proposition 2]). *Let X be a smooth projective variety which is Frobenius split and $\mathcal{L}$ an ample line bundle on X. Then $H^i(X, \mathcal{L}^{-1}) = 0$ for $i < \dim(X)$.*

Proof By Serre duality, $H^i(X, \mathcal{L}^{-m})$ is the dual of $H^{n-i}(X, \omega \otimes \mathcal{L}^m)$ where $n = \dim(X)$. Since $\mathcal{L}$ is ample $H^{n-i}(X, \omega \otimes \mathcal{L}^m)$ vanishes for $i < \dim(X)$ and large m. Thus $H^i(X, \mathcal{L}^{-m})$ vanishes for $i < \dim(X)$ and large m. Now apply Proposition 5.1. $\square$

We continue following [6] for a while. Sometimes we are sketchy. Let $Y \subset X$ be a closed subvariety. We say that $\sigma \in \mathrm{End}_F(X)$ is *compatible* with Y if it maps the ideal sheaf $\mathcal{I}_Y$ of Y to itself. Let $\mathrm{End}_F(X, Y)$ be the set of $\sigma \in \mathrm{End}_F(X)$ compatible with Y. If X has a splitting compatible with Y then we say that Y is *compatibly split*. If a given splitting is compatible with several subvarieties, then we say that these subvarieties are *simultaneously compatibly split*.

Exercise 5.3. *The splitting of $\mathbb{P}^1$ in Exercise 3.3 is compatible with the points 0 and ∞. It corresponds with $\mathrm{dlog}(x)^{1-p} \in \Gamma(\mathbb{P}^1, \omega_{\mathbb{P}^1}^{1-p})$.*

Proposition 5.4 ([6, Proposition 3]). *Let X be a projective variety and $Y \subset X$ a compatibly split closed subvariety. If $\mathcal{L}$ is an ample line bundle on X then the restriction map $H^0(X, \mathcal{L}) \to H^0(Y, \mathcal{L})$ is surjective and $H^i(Y, \mathcal{L})$ vanishes for $i > 0$.*

Sketch of proof Say σ is the compatible splitting. Then $\sigma : F_*\mathcal{O}_X \to \mathcal{O}_X$ is split surjective and induces a split surjective map $F_*\mathcal{I}_Y \to \mathcal{I}_Y$. Arguing as above we get a split surjective map from the $H^i(X, \mathcal{L}^{p^r} \otimes_{\mathcal{O}_X} \mathcal{I}_Y)$ to the $H^i(X, \mathcal{L} \otimes_{\mathcal{O}_X} \mathcal{I}_Y)$. Take r large. $\square$

Lemma 5.5. *Let Y be a closed subvariety of X and let U be an open subset of X such that $U \cap Y$ is dense in Y. Then $\mathcal{E}nd_F(X, Y)$ consists of the $\sigma \in \mathcal{E}nd_F(X)$ whose restriction to U lies in $\mathcal{E}nd_F(U, U \cap Y)$. In other words, compatibility with Y may be checked on U.*

Proof We have already used this principle in Example 2.11. $\square$

Lemma 5.6. *Let Y, Z be closed subvarieties of X and let $\sigma \in \mathcal{E}nd_F(X)$. If σ is compatible with Y and Z, then it is compatible with $Y \cup Z$ and with each* *irreducible component of Y. If σ is compatible with Y and Z, then it is compatible with the scheme theoretic intersection $Y \cap Z$. If Y, Z are simultaneously compatibly split, then their scheme theoretic intersection is reduced.*

Hints for the Proof Say Y is irreducible. By deleting the irreducible components of Z that are not contained in Y one forms an open subset U of X that intersects Y in a dense subset and for which $U \cap (Y \cup Z)$ equals $U \cap Y$. Let V be open. A function $f \in \Gamma(V, \mathcal{O}_X)$ vanishes on $V \cap (Y \cup Z)$ if and only if it vanishes on both $V \cap Y$ and $V \cap Z$. The ideal sheaf of the scheme theoretic intersection $Y \cap Z$ is just $\mathcal{I}_Y + \mathcal{I}_Z$. $\qquad\square$

Proposition 5.7. *Let X be a variety with Frobenius splitting σ. The collection of subvarieties with which σ is compatible is closed under the following operations*

- *Take irreducible components.*

- *Take intersections.*

- *Take unions.* $\qquad\square$

Proposition 5.8 ([6, Proposition 4]). *Let $f : Z \to X$ be a proper morphism of algebraic varieties. Assume that $f_* \mathcal{O}_Z = \mathcal{O}_X$. We then have*

1. *If Z is Frobenius split, then so is X.*

2. *If the closed subvariety Y is compatibly split in Z, then so is $f(Y)$ in X.*

Sketch of proof The idea is that if U is open in X, then a splitting of $\Gamma(U, \mathcal{O}_X)$ amounts to the same as a splitting of $\Gamma(f^{-1}(U), \mathcal{O}_Z)$. $\qquad\square$

Remark 5.9. *[11, Proposition 6.1.6] The condition $f_* \mathcal{O}_Z = \mathcal{O}_X$ is satisfied when f is surjective, proper, has connected fibres, and is separable in the following sense: There is a dense subset of $x \in X$ for which there is $z \in f^{-1}(x)$ for which the tangent map df_z is surjective, where df_z goes from the tangent space at z to the tangent space at x. A birational map is certainly separable.*

5.10. Residues. Let $X = \mathrm{Spec}(A)$ be a smooth affine variety of dimension n and let $f \in A$ have a smooth prime divisor of zeroes $D = \mathrm{div}(f)$. There is a Poincaré residue map $\mathrm{res} : \Gamma(X, \omega_X(D)) \to \Gamma(D, \omega_D)$. Its composite with the surjective map $\beta \mapsto \beta \wedge \mathrm{dlog}(f) : \Gamma(X, \Omega_X^{n-1}) \to \Gamma(X, \omega_X(D))$ is the obvious restriction map $\Gamma(X, \Omega_X^{n-1}) \to \Gamma(D, \omega_D)$. That characterizes the Poincaré residue. We will not actually need the Poincaré residue, but we will use the name *residue* for some related maps, like the map $f^{p-1} * \mathrm{End}_\phi(A) \to \mathrm{End}_\phi(A/(f))$

implied by Lemma 2.8. More specifically, we now seek an explicit formula for the A-linear residue map that is the composite of the maps

$$f^{p-1}\Gamma(X, \omega_X^{1-p}) \cong \mathrm{End}_\phi(A, (f)),$$
$$\mathrm{End}_\phi(A, (f)) \to \mathrm{End}_\phi(A/(f)),$$
$$\mathrm{End}_\phi(A/(f)) \cong \Gamma(D, \omega_D^{1-p}).$$

Lemma 5.11. *Take a point P on D. Let τ be a local generator at P of ω_D and let $\tilde{\tau}$ be a local lift to Ω_X^{n-1}. For any sufficiently small neighborhood U of P the $\Gamma(U, \mathcal{O}_X)$-linear residue map $f^{p-1}\Gamma(U, \omega_X^{1-p}) \to \Gamma(U \cap D, \omega_D^{1-p})$ sends $(\tilde{\tau} \wedge \mathrm{dlog}(f))^{1-p}$ to τ^{1-p}.*

Proof Take U so small that $(\tilde{\tau} \wedge \mathrm{dlog}(f))^{1-p} \in f^{p-1}\Gamma(U, \omega_X^{1-p})$ and unravel the maps. □

Proposition 5.12. *Let X be smooth and let $\sigma \in \Gamma(X, \omega_X^{-1})$ so that σ^{p-1} defines a splitting of X. Then this splitting is compatible with the divisor $\mathrm{div}(\sigma)$ of σ.*

Proof Take a smooth point P on $\mathrm{div}(\sigma)_{\mathrm{red}}$, the reduced subscheme supporting $\mathrm{div}(\sigma)$. Locally around P we are in the situation discussed above: $X = \mathrm{Spec}(A)$, $\sigma \in f^{p-1}\Gamma(X, \omega_X^{-1})$. □

Remark 5.13. *Actually $\mathrm{div}(\sigma)$ must be reduced. If there were a multiple component then one would get a vanishing residue there. Or one could argue that $\mathrm{div}(\sigma)$ is a split scheme, hence reduced.*

Lemma 5.14. *Let X be a smooth projective variety. Let $\sigma \in \mathrm{End}_F(X)$. Suppose $\sigma(1)$ does not vanish at some point $P \in X$. Then σ spans a splitting of X.*

Proof As $\sigma(1) \in \Gamma(X, \mathcal{O}_X)$, it is a constant function. □

Residually normal crossing The Lemma tells that if one has a section σ of $\Gamma(X, \omega_X^{1-p})$, one may check if it spans a splitting by evaluating at a convenient point P. Now it happens often that there is a point P where one may take a residue of σ and thus bring the dimension down. Even better, one may have such luck that by repeatedly taking residues the dimension can be brought all the way down to zero. And then finally, in dimension zero, one hopes to hit a nonzero constant. That yields a rather practical way to establish that σ spans a splitting. The lucky situation we just alluded to has been formalized in [4] with the notion 'residually normal crossing'.

One may find it surprising that residually normal crossings are common. What happens is that in practice σ is not chosen generically but in a very special position so as to have it compatible with an effective divisor that is important in the application at hand. Then residually normal crossing may come as a bonus.

Example 5.15. *We now give an example of the residual normal crossing phenomenon. It is not projective but affine. Actually one may extend the example to nine dimensional projective space, but we will use Exercise 2.4 instead. Let $X = \mathrm{Spec}(A)$ be the coordinate algebra of the space of 3 by 3 matrices. Thus $A = \mathbb{F}[x_{ij}]_{1 \leq i,j \leq 3}$, a polynomial ring in nine variables. Take a volume form $\tau_9 = dx_{11} \wedge dx_{12} \wedge \cdots \wedge dx_{33}$. The 9 refers to the dimension. Note that τ_9^{1-p} does not depend on how we order the variables. Now let $\sigma_9 \in \Gamma(X, \omega_X^{1-p})$ be defined as*

$$\left((x_{11}) \det \begin{pmatrix} x_{11} & x_{12} \\ x_{21} & x_{22} \end{pmatrix} \det \begin{pmatrix} x_{11} & x_{12} & x_{13} \\ x_{21} & x_{22} & x_{23} \\ x_{31} & x_{32} & x_{33} \end{pmatrix} \det \begin{pmatrix} x_{22} & x_{23} \\ x_{32} & x_{33} \end{pmatrix} (x_{33}) \right)^{p-1} \tau_9^{1-p}.$$

Taking a residue at the subvariety $x_{11} = 0$ we get

$$\sigma_8 = \left(\det \begin{pmatrix} 0 & x_{12} \\ x_{21} & x_{22} \end{pmatrix} \det \begin{pmatrix} 0 & x_{12} & x_{13} \\ x_{21} & x_{22} & x_{23} \\ x_{31} & x_{32} & x_{33} \end{pmatrix} \det \begin{pmatrix} x_{22} & x_{23} \\ x_{32} & x_{33} \end{pmatrix} (x_{33}) \right)^{p-1} \tau_8^{1-p}.$$

We can take further residues in several ways, as the function in front of τ_8^{1-p} has several factors of the form x_{ij}^{p-1}. Take residue at the subvariety $x_{12} = 0$, then at its subvariety $x_{21} = 0$. We arrive at

$$\sigma_6 = \left(\det \begin{pmatrix} 0 & 0 & x_{13} \\ 0 & x_{22} & x_{23} \\ x_{31} & x_{32} & x_{33} \end{pmatrix} \det \begin{pmatrix} x_{22} & x_{23} \\ x_{32} & x_{33} \end{pmatrix} (x_{33}) \right)^{p-1} \tau_6^{1-p}.$$

Take residue at $x_{22} = 0$ and one is left with $\sigma_5 = (x_{13} x_{31} x_{23} x_{32} x_{33})^{p-1} \tau_5^{1-p}$. And so on until $\sigma_0 = 1$. What this means is that the original σ_9, viewed as an element of $\mathrm{End}_F(X)$ sends 1 to a function $\sigma_9(1)$ with value 1 at the origin. But σ_9 is also given by a homogeneous formula, so Exercise 2.4 tells that σ_9 defines a splitting. It is compatible with the subvarieties that we encountered along the way. The splitting is also compatible with all the other subvarieties that can be reached in a similar manner, such as the subvariety $x_{11} = x_{33} = 0$. But this is clear from Proposition 5.7 anyway.

Exercise 5.16. *Extend the example to n by n matrices, or even m by n matrices.*

6. Schubert Varieties

It is time to discuss a serious application. Mehta and Ramanathan constructed a Frobenius splitting on the Bott-Samelson-Demazure-Hansen desingularisation of Schubert varieties in a flag variety G/B to show that the Schubert varieties are simultaneously compatibly split in the flag variety. This then leads immediately to the result alluded to in the introduction about intersecting two unions of Schubert varieties. This result about intersections is crucial in the analysis [11] of the fine structure (as B-modules) of dual Weyl modules, nowadays also known as costandard modules $\nabla(\lambda)$. One also immediately gets that if X is a Schubert variety in G/B, and $\mathcal{L}$ is an ample line bundle, then $\Gamma(G/B, \mathcal{L}) \to \Gamma(X, \mathcal{L})$ is surjective and $H^i(G/B, \mathcal{L}) = H^i(X, \mathcal{L}) = 0$ for $i > 0$. However, that is not quite the result that one wants. Kempf vanishing gives that in fact $H^i(G/B, \mathcal{L}) = 0$ for $i > 0$ as soon as $\Gamma(G/B, \mathcal{L}) \neq 0$. For instance, $H^i(G/B, \mathcal{O}_{G/B})$ vanishes for $i > 0$, but $\mathcal{O}_{G/B}$ is not ample. To get a result that covers all of Kempf vanishing, we will need the notion of D-splitting introduced by Ramanan and Ramanathan in [9].

Let us first recall the Bott-Samelson-Demazure-Hansen resolution. We need the usual notations and terminology from the theory of reductive algebraic groups. Let us remind the reader of some of the ingredients in a standard example. See also our book [11] for more details on these constructions.

Example 6.1. *Fix $n > 1$. Let G be the linear algebraic group GL_n over $\mathbb{F}$. As is common, we often discuss things as if G is a group. The true group is $G(\mathbb{F})$, the group of $\mathbb{F}$-rational points of G. By B we denote the algebraic subgroup of upper triangular matrices, by T the algebraic subgroup of diagonal matrices, by $N(T)$ its normalizer, consisting of monomial matrices. (A monomial matrix is invertible and has one nonzero entry in each row.) The Weyl group $W = N(T)/T$ is isomorphic to the symmetric group on n letters. We let S be the set of matrices that can be obtained by permuting two consecutive columns of the identity matrix. One calls S a set of representatives of fundamental reflections in the Weyl group. The number of elements of S is $n-1$. The algebraic subgroup of lower triangular matrices we denote $\tilde{B}$. So B, $\tilde{B}$ are opposite Borel subgroups with intersection T. The flag variety G/B parametrizes flags in n-dimensional vector space. Indeed, an invertible matrix g defines a flag $L_1 \subset L_2 \subset \cdots \subset L_n$ with L_i being the span of the first i columns of g. Matrices g, h define the same flag if and only if the cosets gB, hB are equal. For other reductive groups one still speaks of the flag variety G/B in analogy with this example. For $s \in S$ let P_s be the minimal parabolic subgroup generated by B and s. The subvariety P_s/B of G/B is isomorphic with a projective line $\mathbb{P}^1$. A line bundle $\mathcal{L}$ on G/B is ample if and only if its restriction to P_s/B is ample for each $s \in S$. There is a G-equivariant line bundle $\mathcal{L}_\rho$ on G/B so that $\mathcal{L}_\rho^{-1}$ is 'just ample', meaning that for each $s \in S$ its restriction to P_s/B is the ample generator of the Picard group of P_s/B. One knows ρ as the half sum of the positive roots. A line bundle $\mathcal{L}$ on G/B is ample if and only $\Gamma(G/B, \mathcal{L} \otimes \mathcal{L}_\rho)$ is nonzero. And $\Gamma(G/B, \mathcal{L})$ is nonzero if and only if the $\Gamma(P_s/B, \mathcal{L})$ are nonzero for all $s \in S$.*

Bott-Samelson-Demazure-Hansen resolution   If X, Y are varieties with B acting from the right on X and from the left on Y, then the contracted product $X \times^B Y$ is the the the quotient of $X \times Y$ by the equivalence relation $(xb, y) \approx (x, by)$ for $x \in X$, $y \in Y$, $b \in B$, provided that quotient exists as a variety. If $s_1 s_2 \cdots s_d$ is a word on the alphabet S, so if one is given a sequence of length d with values in S, then we put $Z(s_1 s_2 \cdots s_d) = P_{s_1} \times^B P_{s_2} \times^B \cdots \times^B P_{s_d}/B$. Multiplication defines a map from $Z(s_1 s_2 \cdots s_d)$ to G/B, sending $x_1 \times^B \cdots \times^B x_d B$ to $x_1 \cdots x_d B$. If the word is *reduced* then $Z(s_1 s_2 \cdots s_d) \to G/B$ is birational to its image, which is of dimension d. This may be taken as a definition of *reduced*. The image of $Z(s_1 s_2 \cdots s_d)$ in G/B is the closure of a B-orbit. A B-orbit in G/B is called a *Schubert cell* and its closure is called a *Schubert variety*. Schubert varieties may be singular. The closure of a $\tilde{B}$-orbit is called an *opposite Schubert variety*. The set of Schubert varieties is parametrized by W. If $\dot{w} \in N(T)$ is a representative of $w \in W$, then X_w denotes the closure of the orbit $B\dot{w}B/B$ of $\dot{w}B$. The dimension of X_w is known as the length of w. If $s_1 s_2 \cdots s_d$ is reduced, then it also describes an element $w \in W$ and the birational rational map $Z(s_1 s_2 \cdots s_d) \to X_w$ is known as a Bott-Samelson-Demazure-Hansen resolution of X_w. Indeed, $Z(s_1 s_2 \cdots s_d)$ is smooth: It is an iterated $\mathbb{P}^1$ fibration. The projection map $Z(s_1 s_2 \cdots s_d) \to Z(s_1) = \mathbb{P}^1$, sending $x_1 \times^B \cdots \times^B x_d B$ to $x_1 B$ has fibre $Z(s_2 \cdots s_d)$ above the point B of $Z(s_1)$. We think of B as point zero on this $\mathbb{P}^1$ and we think of $s_1 B$ as the point ∞. On $Z(s_1 s_2 \cdots s_d)$ we have the divisor Z_i consisting of the $x_1 \times^B \cdots \times^B x_d B$ with $x_i = 1$. The divisors $Z_1, \ldots, Z_d$ meet transversely at a point P. If $s_1 s_2 \cdots s_d$ is a reduced word of maximal length, then Mehta and Ramanathan show that $\mathcal{E}nd_F(Z(s_1 s_2 \cdots s_d), Z_1 \cup \cdots \cup Z_d)$ is the pullback from G/B of $\mathcal{L}_\rho^{1-p}$, where $\mathcal{L}_\rho^{-1}$ is the 'just ample' line bundle on G/B. See also [11, Proposition A.4.6], where the same is shown for any word, after Mathieu.

A splitting   The flag variety itself is also a Schubert variety. It corresponds with the longest element w_0 of the Weyl group. Take a Bott-Samelson-Demazure-Hansen resolution $Z(s_1 s_2 \cdots s_d) \to G/B$. (Although it is called a resolution, it is not a resolution of singularities, as G/B itself is smooth.) We wish to take a section $\tau \in \Gamma(G/B, \mathcal{L}_\rho^{-1})$ which does not vanish in the image B of the point P where the Z_i intersect each other. A good choice for τ is a lowest weight vector, or simultaneous eigenvector for $\tilde{B}$, in $\Gamma(G/B, \mathcal{L}_\rho^{-1})$. That works because the $\tilde{B}$-orbit of $B \in G/B$ is dense, so that τ cannot vanish at the point B. We will take τ this way. Let $\sigma \in \mathcal{E}nd_F(Z(s_1 s_2 \cdots s_d), Z_1 \cup \cdots \cup Z_d)$ be the pullback of τ. Then σ^{p-1} spans a splitting because at P there is a residually normal crossing. (A true normal crossing of the Z_i, actually.) This splitting is clearly compatible with the divisor $Z_1 \cup \cdots \cup Z_d$. As G/B is smooth, hence normal, the direct image of the structure sheaf of $Z(s_1 s_2 \cdots s_d)$ must be $\mathcal{O}_{G/B}$,

53

and Proposition 5.8 gives a splitting of G/B compatible with the images of the Z_i. This covers all codimension one Schubert varieties and with Proposition 5.7 one shows that the splitting must be compatible with all Schubert varieties.

Theorem 6.2 (Mehta-Ramanathan). *G/B is Frobenius split with all Schubert varieties compatibly split.*

Remark 6.3. *Mehta and Ramanathan also considered Schubert varieties in G/Q where Q is a parabolic subgroup.*

Normality   We get a nice proof of normality of Schubert varieties by means of the

Lemma 6.4 (Mehta-Srinivas [7]). *Let $f : Y \to X$ be a proper surjective morphism of irreducible $\mathbb{F}$-varieties. Suppose that*

- *Y is normal,*

- *the fibres of f are connected,*

- *X is Frobenius split.*

Then X is normal.

Discussion   The problem is local on X. One argues as in Example 2.10 (the example with the cusp) that if f is in the function field of X so that f^p is a regular function on some open U, then f itself must be a regular function on U. That means that the map from the normalisation of X to X cannot pinch together infinitely near points. In other words, one gets semi-normality in the sense of [10]. As the fibres of f are connected, it is also impossible that disjoint points are pinched. So X is equal to its normalisation. In [1, Proposition 1.2.5] the theme is worked out further by showing that every split scheme X is weakly normal, meaning that every finite birational map $Z \to X$ is an isomorphism.

To apply the Lemma, one could show that a Bott-Samelson-Demazure-Hansen resolution of a Schubert variety has connected fibres, but the argument in [7] is as follows. Let X_w be a Schubert variety in G/B and let $s_1 s_2 \cdots s_d$ be a corresponding reduced word. Let X_z be the image of $Z(s_1 \cdots s_{d-1})$. By induction on dimension we may assume X_z is normal. With the Lemma one shows its image X' in G/P_{s_d} is normal. And the map from X_w to X' is a $\mathbb{P}^1$ fibration. So X_w is normal.

Theorem 6.5. *Schubert varieties are normal.*   □

7. D-splittings

To get more mileage out of the above construction of a splitting on G/B one takes a closer look at τ and $\mathcal{L}_\rho^{-1}$. We have not used yet that $\mathcal{L}_\rho^{-1}$ is ample. The line bundle $\mathcal{L}_\rho^{-1}$ is well understood. Recall that τ is a lowest weight vector in in $\Gamma(G/B, \mathcal{L}_\rho^{-1})$. Its divisor D is the union of the codimension one opposite Schubert varieties. (Compare [11, Exercise 5.2.5].) Our splitting of G/B is thus simultaneously compatible with all Schubert varieties and all opposite Schubert varieties. But let us look at cohomology.

D-splitting If D is an effective divisor then a splitting $F_*\mathcal{O}_X \to \mathcal{O}_X$ of X is called a D-splitting if it factors through the map $F_*\mathcal{O}_X \to F_*(\mathcal{O}_X(D))$. So any D-splitting is a composite $F_*\mathcal{O}_X \to F_*(\mathcal{O}_X(D)) \to \mathcal{O}_X$. If X is smooth, and the section σ of ω_X^{1-p} defines a splitting, then it is a D-splitting precisely if σ lands in the subsheaf $\omega_X^{1-p}(-D)$. For example, in the above construction of the splitting on G/B we may take for D the union of the codimension one opposite Schubert varieties. If X is D-split, then the surjective map $H^i(X, \mathcal{L}^p) \to H^i(X, \mathcal{L})$ factors through $H^i(X, \mathcal{L}^p \otimes \mathcal{O}_X(D))$. So if $i > 0$ and $\mathcal{L}^p \otimes \mathcal{O}_X(D)$ is ample, then it factors through zero by Proposition 5.1. We then conclude that $H^i(X, \mathcal{L})$ vanishes. Thus

Theorem 7.1 (Kempf vanishing). *Let $\mathcal{L}$ be a line bundle on G/B so that $\Gamma(G/B, \mathcal{L})$ is nonzero. Then $H^i(G/B, \mathcal{L})$ vanishes for $i > 0$.*

Proof Indeed, with D as indicated above, $\mathcal{L}^p \otimes \mathcal{O}_X(D) = \mathcal{L}^p \otimes \mathcal{L}_\rho^{-1}$ is ample. $\square$

In similar vein one wants to show

Theorem 7.2. *Let $\mathcal{L}$ be a line bundle on G/B so that $\Gamma(G/B, \mathcal{L})$ is nonzero. Let X_w be a Schubert variety in G/B. Then $\Gamma(G/B, \mathcal{L}) \to \Gamma(X_w, \mathcal{L})$ is surjective and $H^i(X_w, \mathcal{L})$ vanishes for $i > 0$.*

Compatible D-splitting If X is D-split and Y is a subvariety of X then we say that Y is compatibly D-split if Y is compatibly split and no irreducible component of Y is contained in D. Assume this. The complement of D intersects Y in a dense open subset.

We claim that $F_*\mathcal{I}_Y \to \mathcal{I}_Y$ factors through $F_*(\mathcal{I}_Y(D))$. Indeed, $F_*(\mathcal{I}_Y(D)) \to \mathcal{O}_X$ factors through $\mathcal{I}_Y$, because a regular function on an open subset U of X vanishes on $U \cap Y$ if and only if it vanishes on a dense subset of $U \cap Y$.

The surjective map $H^i(X, \mathcal{I}_Y \otimes \mathcal{L}^p) \to H^i(X, \mathcal{I}_Y \otimes \mathcal{L})$ factors through $H^i(X, \mathcal{I}_Y \otimes \mathcal{L}^p \otimes \mathcal{O}_X(D))$, and if $\mathcal{L}^p \otimes \mathcal{O}_X(D)$ is ample this vanishes for $i > 0$, by the proof of Proposition 5.4.

Proof of Theorem 7.2   The Schubert variety is irreducible and contains the point $B \in G/B$ that lies in none of the opposite Schubert varieties. So we may argue as in the proof of Theorem 7.1.   □

8. Canonical Splitting

The group B acts on $\mathrm{End}_F(Z(s_1 s_2 \cdots s_d)) = \Gamma(Z(s_1 s_2 \cdots s_d), \omega^{1-p})$ and one can check that our splitting is given by a T-invariant σ in this B-module. Mathieu has observed that the B-module it generates is rather small. So one might say the splitting is almost B-invariant. Mathieu has formalized this in the notion *canonical splitting* of a variety with B-action.

Recall that a G-module M is called costandard if there is an equivariant line bundle $\mathcal{L}$ on G/B so that $M = \Gamma(G/B, \mathcal{L})$. Mathieu employed canonical splittings to give an amazing proof of the following theorem

Theorem 8.1. *The tensor product of two costandard modules has a filtration by G-submodules whose associated graded module is a direct sum of costandard modules.*

See [11], [5] for an exposition of this.

9. More

There is much more that could be said, but we stop here. The Brion-Kumar book [1] is a treasure trove. If you want to see more recent work, MathSciNet lists over forty references to [1], and Google Scholar lists over a hundred.

References

[1] M. Brion and S. Kumar, Frobenius Splitting Methods in Geometry and Representation Theory, Birkhäuser Boston 2005.

[2] D. Eisenbud, *Commutative algebra. With a view toward algebraic geometry.* Graduate Texts in Mathematics, 150. Springer-Verlag, New York, 1995.

[3] R. Hartshorne, Algebraic Geometry, Graduate Texts in Mathematics 52, Springer-Verlag, Berlin, 1977.

[4] V. Lakshmibai, V.B. Mehta and A.J. Parameswaran, Frobenius splittings and blowups, J. Algebra, 208 (1998), 101–128.

[5] O. Mathieu, Tilting modules and their applications. Analysis on homogeneous spaces and representation theory of Lie groups, Okayama–Kyoto (1997), 145–212, Adv. Stud. Pure Math., 26, Math. Soc. Japan, Tokyo, 2000.

[6] V. B. Mehta and A. Ramanathan, Frobenius splitting and cohomology vanishing for Schubert varieties, Annals of Math. 122 (1985), 27–40.

[7] V.B. Mehta and V. Srinivas, Normality of Schubert varieties, American Journal of Math. 109 (1987), 987–989.

[8] J. Oesterlé, Dégénerescence de la suite spectrale de Hodge vers De Rham, Exposé 673, Séminaire Bourbaki, Astérisque 152–153 (1987), 67–83.

[9] S. Ramanan and A. Ramanathan, Projective normality of flag varieties and Schubert varieties. Invent. Math. 79 (1985), no. 2, 217–224.

[10] R.G. Swan, On seminormality. J. Algebra 67 (1980), no. 1, 210–229.

[11] Wilberd van der Kallen, Lectures on Frobenius splittings and B-modules. Notes by S.P. Inamdar, Tata Institute of Fundamental Research, Bombay, and Springer-Verlag, Berlin, 1993.

On the String Equation of Narasimha

A. S. Vasudeva Murthy*

1. Introduction

In 1968 Roddam Narasimha (RN) published a paper in *Journal of Sound and Vibration* (JSV) deriving the equation

$$\frac{\partial^2 \mathbf{v}}{\partial t^2} + 2R\frac{\partial \mathbf{v}}{\partial t} = \left[1 + \frac{1}{2}\Gamma' \int_0^l \mathbf{v}_x^2 dx \right] \frac{\partial^2 \mathbf{v}}{\partial x^2} + \mathbf{f}_0(x,t) \tag{1.1}$$

for the transverse displacement $\mathbf{v}(x,t)$ of a vibrating string of length l, where $\mathbf{v} = (v, w)$ is a two-dimensional vector in the yz plane orthogonal to the x axis, $\mathbf{v}_x^2 = v_x^2 + w_x^2$ is the squared x derivative of $\mathbf{v}$, R is a damping coefficient, Γ' a nonlinearity parameter that involves a characteristic amplitude of the string motion and the material properties of the string (see equation 4.25 below) and $\mathbf{f}_0$ is an external force acting on the string. Note that for $R \equiv 0$ and $\Gamma' \equiv 0$ this is the standard linear wave equation that is the canonical example for a second order hyperbolic equation in one space dimension which ignores the coupling (usually nonlinear) between the transverse and longitudinal displacements of the string. In the usual text book derivation of the wave equation for the vibrating string it is assumed that the motion of the string is entirely transversal. That this is not strictly true was realized by Kirchhoff [1876] and Lord Rayleigh in the nineteenth century: while the former went on to derive an equation similar to (1.1), Rayleigh [1883] (with no reference to Kirchhoff) restricted himself to the classical van der Pol oscillator to model the vibrating string. More such models were proposed by Osgood in the 1920's, Carrier and Coulson in the 1940's, followed by Oplinger, Murthy and Ramakrishna, Narasimha and Anand in the 1960's. What distinguishes the approach of RN compared to the above mentioned and many others of this period is that RN uses a systematic perturbation expansion of $\mathbf{v}$ in terms of a small parameter ε (a quantity proportional to the amplitude) and then takes the masterly step by assuming that the product of ε^2 with another parameter c_1^2 (which is the

*TIFR-CAM, Bangalore 560065

square of the ratio of longitudinal to transverse wave speeds in the string, generally large in typical metallic wires), is held fixed in the limiting process $\varepsilon \to 0$ (see 4.20 below). This means that higher order equations can also be derived, as RN in fact did. In addition RN derives the exact set of equations by evaluating the acceleration following the particle using the total derivative (as in fluid dynamics), and also taking changes in string length and tension explicitly into account in his equations. This elegant and formal procedure is reflected by Tufillaro's[1989] statement "... if both longitudinal and transverse displacements are small then the simplest model of a string is necessarily nonlinear. This point is demonstrated by Narasimha". Watzky [1992] points out that it was only RN and Nayfeh and Mook [1979] (who refer to RN) who displayed conditions under which the contribution of deflection to the extension of the string leads to transverse vibrations. In fact, Nayfeh and Mook [1979, page 308 and 310] state that RN presents a careful development of the nonlinear equations of motion and a thorough discussion of the various assumptions involved. This was essential because of the controversy surrounding the assumption of zero longitudinal displacement in the previous studies. His meticulous approach helps RN to deduce that string motions with such an assumption are generally not possible. Leamy and Gottlieb [2001], while deriving a 3D model for the vibrating string that includes nonlinear effects in the material law, point out that although RN uses a linear material description they are general enough to include nonlinearities but the equations thus obtained are more complex than theirs. Further, they point out that among the studies that use linear material laws, RN's model (for transverse vibrations) correctly accounts for nonresonant longitudinal motion. Finally, Bisen [2007], in his PhD thesis, states that it was RN who pointed out that, even under small amplitude motion, the equations are inherently nonlinear and RN showed that neglecting axial stretch and changes in tension along the length of the string is not justifiable. Bisen [2007] then goes on to state that while this was implicit in the earlier analysis (Routh, 1905) it was RN who made the representation of length and tension changes explicit and showed further that planar motion was unstable at a high enough amplitude.

It was perhaps RN's rigorous derivation that prompted the great Euler scholar Clifford Truesdell, and a well known authority on the history of the mathematical foundations of continuum mechanics, to write to RN personally (cf Narasimha 1969) regarding the use of material variables by Euler prior to Lagrange, to whom RN (and others) attributed the idea. RN's derivation preceded by more than a decade the lament of Antman [1980] that an honest derivation of the classical equations of motion for strings was lacking in basic books on partial differential equations. As pointed out by Yong [2006], this state of affairs continues to be true (save for a few like Pinsky 1998 and Kevorkian 2000) even three decades after Antman's book, and therefore it is important that students be introduced to the wave equation in a flexible, physically realistic and systematic way: he calls this an "honest" way of deriving the wave equation. Also RN's work revived mathematical studies of (1.1) after Lions [1978] formulated it in an abstract framework and proved the existence and uniqueness of the solution. As Villaggio [1997, p289] remarks, the mathematical treatment of (1.1) is very recent. We say more on this in the Discussion section below.

2. History

The standard linear wave equation

$$\frac{\partial^2 v}{\partial t^2} = \frac{\partial^2 v}{\partial x^2}$$

was derived first by d'Alembert in 1747 and Euler in 1748. Daniel Bernoulli gave a solution in terms of sinusoidal series in 1753. In 1759 Lagrange derived a spatially discretized version of the wave equation. During this period there was intense debate regarding the solution of the wave equation, as the notion and validity of Fourier series was yet to be established (this was done in 1807, see Wheeler and Crummet 1987).

The following account of Kirchhoff's derivation is based on Villaggio [1997]. **Kirchhoff [1876]** considered planar oscillations in the x and z plane of an infinitely thin rod (*eines unendlich dünnen Stabes*), with the respective displacements by $(u(x,t), w(x,t))$. He then applies Hamilton's principle between times t_1 and t_2,

$$\delta \int_{t_1}^{t_2} (W - T)(t)dt = 0$$

where T, W are the total kinetic and strain energy respectively. Here T is given by

$$T = \frac{1}{2}\rho A \int_0^l [u_t^2 + w_t^2]dx$$

where ρ denotes the density of material and A is the cross sectional area of the rod that was assumed to be prismatic. The total strain energy is expressed as

$$W = \int_0^l \phi(x)dx$$

where ϕ is the strain energy per unit length taken to be

$$\phi = \frac{1}{2}S[u_x + \frac{1}{2}w_x^2]^2 + \frac{1}{2}Bw_{xx}^2,$$

here S and B are respectively the extensional and flexural rigidity. The longitudinal strain (expressed in terms of displacement) in the x direction is

$$\varepsilon_x = \sqrt{1 + \lambda_{xx}} - 1$$

where

$$\lambda_{xx} = 2u_x + u_x^2 + w_x^2.$$

Using a binomial expansion ignoring u_x^2 but retaining w_x^2 Kirchhoff obtains a first order approximation

$$\varepsilon_{x0} = \frac{1}{2}\lambda_{xx} = u_x + \frac{1}{2}w_x^2.$$

Using these approximations in the above Hamilton's principle and equating to zero the coefficients of δu and δw under the sign of double integration, Kirchhoff obtained the Euler equations for the variational equation:

$$S\frac{\partial}{\partial x}\left[\frac{\partial u}{\partial x} + \frac{1}{2}\left(\frac{\partial w}{\partial x}\right)^2\right] = \rho A \frac{\partial^2 u}{\partial t^2}$$

$$-B\frac{\partial^4 w}{\partial x^4} + S\frac{\partial}{\partial x}\left[\frac{\partial w}{\partial x}\left(\frac{\partial u}{\partial x} + \frac{1}{2}\left[\frac{\partial w}{\partial x}\right]^2\right)\right] = \rho A \frac{\partial^2 w}{\partial t^2}.$$

To obtain a single equation Kirchhoff then neglects u_{tt} (which means longitudinal waves are ignored or considered very fast) in the first equation. This yields

$$\frac{\partial u}{\partial x}(x,t) + \frac{1}{2}\left[\frac{\partial w}{\partial x}(x,t)\right]^2 = \sigma(t) \tag{2.2}$$

where σ can be determined by integrating (2.2) with respect to x

$$\sigma(t) = \frac{1}{l}\int_0^l\left[\frac{\partial u}{\partial x} + \frac{1}{2}\left(\frac{\partial w}{\partial x}\right)^2\right]dx = \frac{1}{l}\left[u(l,t) - u(0,t) + \frac{1}{2}\int_0^l w_x^2 dx\right].$$

Using this in the equation for w we obtain

$$\rho A\frac{\partial^2 w}{\partial t^2} + B\frac{\partial^4 w}{\partial x^4} = \frac{S}{l}\left[u(l,t) - u(0,t) + \frac{1}{2}\int_0^l w_x^2 dx\right]\frac{\partial^2 w}{\partial x^2}.$$

Setting $B = 0$ (i.e. ignoring bending), and using the boundary conditions $u(0,t) = 0, u(l,t) = u_l$ we obtain

$$\rho A\frac{\partial^2 w}{\partial t^2} = \frac{S}{l}\left[u_l + \frac{1}{2}\int_0^l w_x^2 dx\right]\frac{\partial^2 w}{\partial x^2}.$$

This is the equation (23) on page 444 of Kirchhoff [1876] which (in his notation) reads

$$\frac{\mu}{E}\frac{\partial^2 \xi}{\partial t^2} = \left(\frac{\omega'}{l} + \frac{1}{2l}\int_0^l \left(\frac{\partial \xi}{\partial s}\right)^2 ds\right)\frac{\partial^2 \xi}{\partial s^2}.$$

As stated by Arosio [1995] - "The original deduction of Kirchhoff was obtained under the idea of looking only for motions where $u_{tt} = o(w_{tt})$ (*Wir wollen nur solche Bewegungen aufsuchen, bei welchen u_{tt} ist unendlich klein gegen w_{tt}*). This led him to the conclusion that for such a motion the strain ε_{x0} must satisfy the condition $\varepsilon_{x0} \approx$ independent of x." However RN obtains this approximation as a leading order term in his expansion of λ, see (4.12), (4.16)-(4.17) and (4.21) below.

Osgood (1925) proceeds in a different way. Considering an infinitesimal length of a perfectly elastic string of density ρ with its two ends fixed Osgood derived the equations for the longitudinal displacement u and transverse displacement v

$$\rho\frac{\partial^2 u}{\partial t^2} = \frac{\partial}{\partial x}\left[cu_x - \frac{\lambda(1+u_x)}{\sqrt{(1+u_x)^2 + v_x^2}}\right]$$

$$\rho\frac{\partial^2 v}{\partial t^2} = \frac{\partial}{\partial x}\left[cv_x - \frac{\lambda v_x}{\sqrt{(1+u_x)^2 + v_x^2}}\right]$$

by writing down Newton's second law

$$\rho\frac{\partial^2 u}{\partial t^2} = \frac{\partial}{\partial x}(T\cos\theta)$$

$$\rho\frac{\partial^2 v}{\partial t^2} = \frac{\partial}{\partial x}(T\sin\theta)$$

(2.3)

where T is the tension acting along the tangent of the deflected (at local angle θ to the x-axis) string. Using Hooke's law

$$T = (T_0 + \lambda)\frac{\partial s}{\partial x} - \lambda$$

where T_0 is the tension at equilibrium, λ is now the "spring" constant and s is the arc length

$$\frac{\partial s}{\partial x} = \sqrt{(1+u_x)^2 + v_x^2}.$$

Using the expressions

$$\sin\theta = \frac{v_x}{\sqrt{(1+u_x)^2 + v_x^2}}$$

$$\cos\theta = \frac{1+u_x}{\sqrt{(1+u_x)^2 + v_x^2}}$$

Osgood obtains the above equations with $c = T_0 + \lambda$.

In 1941 Coulson (see Coulson and Jefferey 1977) derived the equation

$$\frac{\partial^2 v}{\partial t^2} = \frac{c^2}{[1+v_x^2]^2} \frac{\partial^2 v}{\partial x^2}.$$

Proceeding along the lines of Osgood but considering only transverse displacement and assuming that the tangents at the end points of the infinitesimal length ds of the string to be θ and $\theta + d\theta$, we have

$$\rho ds \frac{\partial^2 v}{\partial t^2} = T_0 \sin(\theta + d\theta) - T_0 \sin\theta.$$

Neglecting higher order terms

$$\rho ds \frac{\partial^2 v}{\partial t^2} = T_0 \cos\theta d\theta.$$

Using $\tan\theta = v_x$ Coulson obtains

$$\frac{\partial^2 v}{\partial t^2} = T_0/\rho \frac{\cos\theta}{\sec^2\theta} \frac{dx}{ds} \frac{\partial^2 v}{\partial x^2} = \frac{c^2}{\sec^4\theta} \frac{\partial^2 v}{\partial x^2} = \frac{c^2}{[1+v_x^2]^2} \frac{\partial^2 v}{\partial x^2}$$

where $c^2 = T_0/\rho$.

Carrier [1945, 1949] obtained the equation

$$\frac{\partial^2 \varphi}{\partial t^2} = [1 + \frac{1}{2\pi} \int_0^l \varphi^2(x,t)dx] \frac{\partial^2 \varphi}{\partial x^2}$$

by considering Newton's equations (2.3) and postulating the stress strain relationship

$$T = T_0 EA \left[\sqrt{(1+v_x)^2 + u_x^2} - 1 \right]$$

63

where T_0 is the rest tension, E is an elastic constant of the material and A the rest cross-sectional area. With zero boundary conditions for u and eliminating u_x, Carrier obtains the equation for

$$\varphi = \frac{v_x}{1 + \alpha^2 \tau}$$

where

$$\alpha^2 = T_0/EA, \quad \tau = \frac{T - T_0}{T_0}$$

Nondimensionalizing $\bar{x} = \pi x/l$ and $\bar{t} = t\frac{\pi}{e}\sqrt{\frac{T_0}{\rho A}}$ Carrier obtains

$$\frac{\partial^2}{\partial \bar{t}^2}\left[(1 + \alpha^2 \tau)\varphi\right] = \frac{\partial^2}{\partial \bar{x}^2}\left[(1 + \tau)\varphi\right]$$

$$\frac{\partial^2}{\partial \bar{t}^2}\left[(1 + \alpha^2 \tau)\psi\right] = \frac{\partial^2}{\partial \bar{x}^2}\left[(1 + \tau)\psi\right]$$

where

$$\psi = \sqrt{1 - \varphi^2}.$$

The boundary conditions are equivalently written as

$$\int_0^\pi (1 + \alpha^2 \tau)\psi d\bar{x} = \pi.$$

Carrier then argues that the α^2 terms can be neglected to obtain

$$\frac{\partial^2 \varphi}{\partial \bar{t}^2} = \frac{\partial^2}{\partial \bar{x}^2}\left[(1 + \tau)\varphi\right] \qquad (2.4)$$

$$\frac{\partial^2 \psi}{\partial \bar{t}^2} = \frac{\partial^2}{\partial \bar{x}^2}\left[(1 + \tau)\psi\right]$$

and approximating

$$\psi \approx 1 - \frac{1}{2}\varphi^2$$

leading to

$$\int_0^\pi \left[1 + \alpha^2 \tau(\bar{t})\right]\left[1 - \frac{1}{2}\varphi^2\right] d\bar{x} = \pi.$$

Carrier then ignores the term $\alpha^2 \tau(\bar{t})\varphi^2$ to obtain

$$\tau = \frac{1}{2\pi\alpha^2}\int_0^\pi \varphi^2 d\bar{x}$$

64

which when substituted in (2.4) leads to

$$\frac{\partial^2 \varphi}{\partial \bar{t}^2} = \left[1 + \frac{1}{2\pi\alpha^2} \int_0^\pi \varphi^2 \, d\bar{x} \right] \frac{\partial^2 \varphi}{\partial \bar{x}^2}.$$

Subsequently in 1949 Carrier modified the above derivation by taking $\bar{t} = (T_0/\rho A)^{1/2} t\pi/l$ but retaining the other variables in Newton's equation which written in the complex form

$$\frac{\partial^2}{\partial \bar{\tau}^2}(1 + \alpha^2 \tau)e^{i\theta} = \frac{\partial^2}{\partial \bar{x}^2}(1 + \tau)e^{i\theta}$$

leads to the integral condition

$$\int_0^\pi (1 + \alpha^2 \tau)e^{i\theta} \, d\bar{x} = \pi.$$

Note that

$$\sin\theta = \frac{v_x}{1 + \alpha^2 \tau}, \quad \cos\theta = \frac{1 + u_x}{1 + \alpha^2 \tau}$$

where

$$1 + \alpha^2 \tau = \sqrt{(1 + v_x)^2 + u_x^2}.$$

Carrier then sets $\theta = \alpha w$, which means $\alpha w \approx v_x/(1 + \alpha^2 \tau)$, to obtain

$$\frac{\partial^2 w}{\partial \bar{t}^2} - (1 + \tau)\frac{\partial^2 w}{\partial \bar{x}^2} = 2w_{\bar{x}}\tau_{\bar{x}} - \alpha^2 \left[2w_{\bar{t}}\tau_{\bar{t}} + \tau w_{\bar{t}\bar{t}} \right]$$

$$\frac{\partial^2 \tau}{\partial \bar{x}^2} = \alpha^2 \left[\frac{\partial^2 \tau}{\partial \bar{t}^2} + (1 + \tau)w_{\bar{x}}^2 - w_{\bar{t}}^2 \right] - \alpha^4 \tau w_{\bar{t}}^2$$

with the integral condition

$$\int_0^\pi \left[\left\{ \tau - \frac{w^2}{2} \right\} + \alpha^2 \left\{ \frac{w^4}{4!} - \tau \frac{w^2}{2} \right\} + \ldots \right] d\bar{x} = 0.$$

For $\alpha^2 = 0$ we obtain

$$\tau = \frac{1}{2\pi} \int_0^\pi w^2 \, d\bar{x}$$

which leads to

$$\frac{\partial^2 w}{\partial \bar{t}^2} = \left[1 + \frac{1}{2\pi} \int_0^\pi w^2 \, d\bar{x} \right] \frac{\partial^2 w}{\partial \bar{x}^2}.$$

Kurmyshev [2003] argues that the mathematical results of Carrier's equation is difficult to compare with those obtained experimentally as the variables of the equation are tangent vectors and string tension, the latter is not measurable for thin strings. In addition, in the first approximation the tension variable depends only on time and has no spatial dependence hence the strain is uniform along the string at any instant which can only be valid for metallic strings. Thus Carrier's model cannot be valid for rubber or soft nylon strings.

Ficken [1957] modified Osgood's equations to

$$\rho\frac{\partial^2 u}{\partial t^2} = (\tilde{T} + \frac{\lambda(1+u_x)}{s_x^2})\frac{\partial^2 u}{\partial x^2}$$

$$\rho\frac{\partial^2 v}{\partial t^2} = \tilde{T}\frac{\partial^2 v}{\partial x^2} + \frac{\lambda v_x}{s_x^2}\frac{\partial^2 u}{\partial x^2}$$

where $\tilde{T} = T_0 + \lambda/u_x$ and s_x and λ are as in Osgood's equations. It was obtained by introducing the so called "slightly elongated" and "slightly inclined" approximations

$$|T - T_0| << T_0$$

$$\sin^2\theta << 1$$

or

$$v_x^2 << s_x^2.$$

Ficken also showed that his equations can be obtained as Euler's equations for minimizing an appropriate integral.

Lee[1957] considered the equation for v given in (2.3) (with $u = 0$) and then approximates T by

$$T = T_0 + \frac{1}{2}AEv_x^2$$

to obtain (after expressing $\sin\theta$ in terms of v_x)

$$\rho A\frac{\partial^2 v}{\partial t^2} = \left[T_0 + \frac{3}{2}AEv_x^2\right]\frac{\partial^2 v}{\partial x^2}.$$

Oplinger [1960] begins with Carrier's equation for transverse motions (cf equation for v in 2.3) and then makes the approximation (small-amplitude motion)

$$v_x \approx \theta \approx \sin\theta$$

66

leading to

$$T = T_0 + \frac{1}{2}\frac{\lambda}{l}\int_0^l v_x^2 dx$$

thus obtaining Kirchhoff's equation

$$\rho\frac{\partial^2 v}{\partial t^2} = \left[T_0 + \frac{1}{2}\frac{\lambda}{l}\int_0^l v_x^2 dx\right]\frac{\partial^2 v}{\partial x^2}.$$

Note that this is different from Carrier's equation as the latter does not contain the derivative term in the integral.

Miles [1965] does not derive a PDE but instead derives an ODE by substituting the Fourier series in the Lagrangian

$$L = K - V$$

where K is the kinetic energy

$$K = \frac{1}{2}\rho\int_0^l (v_t^2 + w_t^2)dx$$

and V is the potential energy

$$V = \frac{1}{2}E\int_0^l [s_0\lambda^2 + \frac{1}{4}\lambda^4 + O(\lambda^6)]dx$$

where E is the elastic constant, s_0 is the uniform strain in the equilibrium position and $\lambda = \sqrt{v_x^2 + w_x^2}$. Considering only the first term in the Fourier series Miles obtains the ODE

$$\frac{1}{\delta}(1 + D^2) + \frac{2}{3}(\alpha^2 + \beta^2)\alpha = \cos(\omega t)$$

$$\frac{1}{\delta}(1 + D^2) + \frac{2}{3}(\alpha^2 + \beta^2)\beta = 0$$

where δ is a small parameter and $D = p^{-1}d/dt$

Murthy and Ramakrishna [1965] applied the Hamilton's principle to the above Lagrangian L and obtained the equations (from the minimizing functions that satisfy the Euler's equations)

$$\frac{\partial^2 u}{\partial t^2} - c_0^2\frac{\partial^2 u}{\partial x^2} - \frac{3}{2}c_1^2 u_{xx}u_x^2 - \frac{1}{2}c_1^2\frac{\partial}{\partial x}\left(u_x w_x^2\right) = -\frac{1}{m}f(x)\cos\omega t$$

(2.5)

$$\frac{\partial^2 w}{\partial t^2} - c_0^2\frac{\partial^2 \omega}{\partial x^2} - \frac{3}{2}c_1^2 w_{xx}w_x^2 - \frac{1}{2}c_1^2\frac{\partial}{\partial x}(u_x^2 w_x) = 0$$

67

where c_0 and c_1 are the transverse and longitudinal velocities of the string, and $f(x) \cos \omega t$ is the externally applied force acting in the y direction. Both Miles[1965] and Murthy and Ramakrishna [1965] neglected the coupling between the longitudinal and transverse modes of oscillation. Towards this end Miles [1984] points out how RN showed that longitudinal displacement of a particle (which Miles 1965 had neglected) in the string must be included in a consistent formulation.

Anand [1966] approximates T (after including the w component as well) in the stress strain relationship of Carrier to obtain

$$T = T_0 + EA \left(u_x + \frac{1}{2} v_x^2 + \frac{1}{2} w_x^2 \right)$$

ignoring third and higher order terms in the arc length obtaining

$$\frac{\partial^2 u}{\partial t^2} = c_1^2 \frac{\partial^2 u}{\partial x^2} + \frac{1}{2}(c_1^2 - c_0^2) \frac{\partial}{\partial x}(v_x^2 + w_x^2)$$

where

$$c_0 = \sqrt{T_0/\rho}, \quad c_1 = \sqrt{EA/\rho}.$$

Arguing that transverse vibration of the string is also subjected to a viscous damping force $2R$, retaining second order terms in v_x and w_x, but to first order. for u_x in T and the arc length, Anand obtains

$$\frac{\partial^2 v}{\partial t^2} + 2R \frac{\partial v}{\partial t} = c_0^2 \frac{\partial^2 v}{\partial x^2} + (c_1^2 - c_0^2) \frac{\partial}{\partial x} v_x \left(u_x + \frac{1}{2} v_x^2 + \frac{1}{2} w_x^2 \right)$$

$$\frac{\partial^2 w}{\partial t^2} + 2R \frac{\partial w}{\partial t} = c_0^2 \frac{\partial^2 v}{\partial x^2} + (c_1^2 - c_0^2) \frac{\partial}{\partial x} w_x \left(u_x + \frac{1}{2} v_x^2 + \frac{1}{2} w_x^2 \right).$$

Anand then goes on to examine typical values for c_1^2/c_0^2 and finds that they lie in the range 400-1000 for metallic strings and hence replaces $c_1^2 - c_0^2$ by c_1^2 and then arguing that the last term in the above equation for u maybe thought of as a driving force for the longitudinal vibrations and that the frequency spectrum of this driving force will be well below the lowest longitudinal resonance frequency one can therefore neglect u_{tt} to get

$$u_x = -\frac{1}{2}(v_x^2 + w_x^2) + \frac{1}{2l} \int_0^l (v_x^2 + w_x^2) dx$$

and using zero boundary conditions at 0 and l

$$u(x,t) = -\frac{1}{2} \int_0^x (v_x^2 + w_x^2) dx + \frac{x}{2l} \int_0^l (v_x^2 + w_x^2) dx$$

68

which then yields the equations

$$\frac{\partial^2 v}{\partial t^2} + 2R\frac{\partial v}{\partial t} = c_0^2\frac{\partial^2 v}{\partial x^2} + \frac{c_1^2}{2l}v_{xx}\int_0^l (v_x^2 + w_x^2)dx$$

$$\frac{\partial^2 w}{\partial t^2} + 2R\frac{\partial w}{\partial t} = c_0^2\frac{\partial^2 w}{\partial x^2} + \frac{c_1^2}{2l}w_{xx}\int_0^l (v_x^2 + w_x^2)dx.$$

Anand then gets a solution in Fourier series for sinusoidal initial conditions.

Stuart [1966] proceeding along the lines of Carrier by considering Newton's equations and the stress-strain relationship but assuming that displacements may be separated into products of functions of position and of time (separation of variables) with a factor $\mathcal{A}$ treated as an amplitude factor. Carrying out the differentiations Stuart obtains

$$\frac{\partial^2 u}{\partial \underline{t}^2} = \frac{\partial^2 u}{\partial \bar{x}^2} + \frac{1}{e}\frac{\partial}{\partial \bar{x}}[u_{\bar{x}}(v_{\bar{x}} + \frac{1}{2}u_{\bar{x}}^2)] + O(\mathcal{A}^5)$$

$$e\frac{\partial^2 v}{\partial \underline{t}^2} = (1 + e)\frac{\partial^2 v}{\partial \bar{x}^2} + u_{\bar{x}}u_{\bar{x}\bar{x}} + O(\mathcal{A}^4)$$

where

$$e = \frac{T_0}{EA - T_0}, \quad \underline{t} = \frac{\pi}{l}\sqrt{\frac{T_0}{\rho}}, \quad \bar{x} = x/l.$$

Analyzing the above Stuart concludes that they are consistent with that of Carrier.

Thus we see that there is no single equation that is common to all the authors. Moreover they are not derived from a single basic equation based on order of parameters or systematic perturbation analysis as done by RN. The Table in the next page summarises this fact.

3. Classical Derivation of the Wave Equation

The instantaneous state of a deformable string is described by a time dependent displacement which indicates how much the string at time t is displaced from its original position x. We assume that the string of length l is at rest occupying the interval $(0, l)$ on the x - axis. The longitudinal displacement of a particle on the string at x and time t is denoted by $u(x, t)$ and $(v, w)(x, t)$ is the transverse displacement. The displacement vector is denoted by

$$U(x, t) = [u(x, t), v(x, t), w(x, t)]^t$$

Author	Longitudinal displacement	Lateral displacement	Total derivative	Damping term	Integral term	Remarks
Kirchhoff 1876	Yes	No	No	No	Yes	Nonlocal nonlinear (in v_x) PDE
Rayleigh 1883	No	No	No	Yes	No	ODE
Osgood 1925	Yes	No	No	No	No	Coupled nonlinear PDE
Coulson 1941	No	No	No	No	No	Single nonlinear PDE
Carrier 1945	Yes	No	No	No	Yes	Nonlocal nonlinear (in v) PDE with amplitude as the small parameter
Ficken 1957	Yes	No	No	No	No	Coupled nonlinear PDE
Lee 1957	No	No	No	No	No	Single nonlinear PDE
Oplinger 1960	Yes	No	No	No	Yes	Nonlocal nonlinear PDE
Murthy & Ramakrishna 1964	No	Yes	No	No	No	Coupled nonlinear PDE
Miles 1965	No	Yes	No	No	No	ODE
Anand 1966	Yes	Yes	No	Yes	Yes	Coupled nonlinear PDE
Stuart 1966	Yes	No	No	No	No	Coupled nonlinear PDE
Narasimha 1968	Yes	Yes	Yes	Yes	Yes	Nonlocal nonlinear (in v_x)PDE with amplitude as the small parameter and ratio of speeds as the large parameter

and its velocity
$$V(x,t) = [u_t(x,t), v_t(x,t), w_t(x,t)]^t.$$
We assume that $U(\cdot, t)$ and $V(\cdot, t)$ are smooth maps from $[0, l] \rightarrow \mathbb{R}^3$. The cross-section of the string can vary when stretched, consequently the density or mass can vary with x and t. We denote $m(x,t)$ to be the mass per unit length of the string. By balancing forces on an arbitrary segment lying between x_1 and x_2, we obtain

$$\frac{d}{dt} \int_{x_1}^{x_2} m(s,t)V(s,t)ds = T(x_2, t) - T(x_1, t) + \int_{x_1}^{x_2} F(s,t)ds \qquad (3.6)$$

Here $T(x,t)$ represents the force experienced by the segment to the left of x due to the segment right of x; similarly the force that the segment right of x experiences is $-T(x,t)$. Assuming the string to be of constant mass in space and time

$$m \int_{x_1}^{x_2} \frac{\partial V}{\partial t}(s,t)ds = \int_{x_1}^{x_2} \frac{\partial T}{\partial x}(s,t)ds + \int_{x_1}^{x_2} F(s,t)ds.$$

Differentiating with respect to x_2 (or x_1) we obtain

$$m \frac{\partial V}{\partial t}(x,t) = \frac{\partial T}{\partial x}(x,t) + F(x,t). \qquad (3.7)$$

To complete the equation we need a model for T. Since tension is a contact force and for strings it acts in the direction of the tangent to the curve of the deflected string at x, we can take

$$T(x,t) = \widetilde{T}(x,t) \ U_x(x,t)/\|U_x\| \qquad (3.8)$$

where $\|.\|$ is the standard Euclidean norm and $\widetilde{T}(x,t)$ is the tension multiplying the unit tangent vector. For a perfectly elastic material

$$\widetilde{T} = E\|U_x\| \qquad (3.9)$$

where E is the Young's modulus. Using (3.8)–(3.9) in (3.7) we obtain

$$m \frac{\partial^2 u}{\partial t^2}(x,t) = E \frac{\partial^2 u}{\partial x^2}(x,t) + F_1(x,t)$$

$$m \frac{\partial^2 v}{\partial t^2}(x,t) = E \frac{\partial^2 v}{\partial x^2}(x,t) + F_2(x,t)$$

$$m \frac{\partial^2 w}{\partial t^2}(x,t) = E \frac{\partial^2 w}{\partial x^2}(x,t) + F_3(x,t)$$

which was derived by Keller (1959).

71

4. Narasimha's Derivation

RN's main point is to assume that the string is the limit of a long elastic rod and the motion of a particle on the string is not confined to a straight line. Therefore the velocity of motion must be defined by the total derivative taking care that acceleration must be evaluated following the particle. Towards this RN defines the material position or the Lagrangian coordinate

$$\xi(x,t) = x + u(x,t)$$

and the inverse by

$$x(\xi,t) = \xi - u(\xi,t)$$

Now consider an arbitrary segment lying between (ξ_1, ξ_2) and rewrite (3.6) as

$$\frac{d}{dt} \int_{\xi_1}^{\xi_2} m(s,t)v(s,t)ds = T(\xi_2,t) - T(\xi_1,t) + \int_{\xi_1}^{\xi_2} F(s,t)ds.$$

Since ξ is time dependent we obtain from the Reynold's transport theorem

$$\int_{\xi_1}^{\xi_2} \left(\frac{\partial}{\partial t}(mV) + \frac{\partial}{\partial x}(mV^2) \right)(s,t)ds = \int_{\xi_1}^{\xi_2} \frac{\partial T}{\partial x}(s,t)ds + \int_{\xi_1}^{\xi_2} F(s,t)ds$$

where $V^2 = [u_t^2, v_t^2, w_t^2]^t$. Now RN uses the conservation of mass

$$\frac{\partial m}{\partial t} + \frac{\partial}{\partial x}(mV) = 0$$

which gives

$$\int_{\xi_1}^{\xi_2} m(s,t)\frac{DV}{Dt}(s,t)ds = \int_{\xi_1}^{\xi_2} \left[\frac{\partial T}{\partial x}(s,t) + F(s,t) \right] ds$$

where

$$\frac{D}{Dt} = \frac{\partial}{\partial t} + u\frac{\partial}{\partial x} \tag{4.10}$$

$$m(x,t)\frac{DV}{Dt}(x,t) = \frac{\partial T}{\partial x}(x,t) + F(x,t).$$

Now conservation of mass gives

$$\int_{x_1}^{x_2} m_0(s)ds = \int_{\xi_1}^{\xi_2} m(s,t)ds$$

where $m_0(s)$ is the initial mass distribution (the value of m when the string is at rest). Changing variables gives

$$\int_{\xi_1}^{\xi_2} m_0(s - u(s,t))(1 - u_s(s,t))ds = \int_{\xi_1}^{\xi_2} m(s,t)ds,$$

yielding

$$(1 - u_x)m_0(x - u)\frac{DU_t}{Dt} = \frac{\partial T}{\partial x} + F. \tag{4.11}$$

The forcing F can be split as

$$F(x,t) = [h(x,t), f_1(x,t) + g_1(x,t), f_2(x,t) + g_2(x,t)]^t = [h, \mathbf{f} + \mathbf{g}]^t$$

where $\mathbf{f}$ is an external force acting on the string in the plane normal to x and h is a similar forcing along the longitudinal direction while $\mathbf{g}$ is the damping force. The system (4.11) derived by RN are the exact equations of motion for a string and are more general than Antman (1980) (see also Yong 2006) as the factor $1 - u_x$ obtained by taking the convective (or, in this case, advective) derivative does not appear in Antman's equation.

To model the relationship between T and the displacements RN assumes that the string is a linear elastic solid very thin compared to its length (so that stress may be taken to be uniform across its length, which means bending and rigidity modulus are ignored). Consequently the tension at any point on the string is simply the product of the Young's modulus E of the material, the cross section area A and the local strain e. With these assumptions the tension T can be written as

$$T = \begin{pmatrix} T_1 \\ T_2 \\ T_3 \end{pmatrix} = \begin{pmatrix} \Lambda \\ \Lambda v_x \\ \Lambda w_x \end{pmatrix} = \Lambda \begin{pmatrix} 1 \\ v_x \\ w_x \end{pmatrix} \tag{4.12}$$

where

$$\Lambda = \frac{1 + c_1^2 \lambda - c_2^2 \lambda^2 + c_3^2 \lambda^3}{c_1^2(1 + \lambda)(1 - u_x)}, \tag{4.13}$$

$$\lambda = \frac{\sqrt{1 + v_x^2 + w_x^2}}{(1 - u_x)} - 1, \tag{4.14}$$

and

$$c_1^2 = \frac{1}{\lambda_0} - 2\nu',$$

$$c_2^2 = \frac{2\nu'}{\lambda_0} - \nu'^2,$$

$$c_3^2 = \nu'^2/\lambda_0,$$

$$\lambda_0 = \frac{e_0}{1 + e_0},$$

$$\nu' = \frac{\nu(1 + e_0)}{(1 - \nu e_0)}.$$

Here e_0 is the initial strain corresponding to $T_0 \equiv T(x, 0)$, and ν is the Poisson ratio which takes into account the lateral contraction when the string is stretched. Assuming m_0 to be constant i.e independent of x the system (4.11) reduces to (for $\mathbf{f} = \mathbf{0}$)

$$(1 - u_x)\frac{D}{Dt}u_t = c_1^2 \frac{\partial}{\partial x}T_1 + h$$

$$(1 - u_x)\frac{D}{Dt}v_t = c_1^2 \frac{\partial}{\partial x}(T_2) + f_1 + g_1 \qquad (4.15)$$

$$(1 - u_x)\frac{D}{Dt}w_t = c_1^2 \frac{\partial}{\partial x}(T_3) + f_2 + g_2.$$

A simple geometric argument shows that for small transverse amplitudes (δ) the longitudinal amplitude is $O(\delta^2)$. So one can assume, for small amplitude displacements, the initial conditions

$$u(x, 0) = \varepsilon^2 a_0(x), \quad v(x, 0) = \varepsilon b_0(x), \quad w(x, 0) = \varepsilon d_0(x),$$

$$u_t(x, 0) = \varepsilon^2 a_1(x), \quad v_t(x, 0) = \varepsilon b_1(x), \quad w_t(x, 0) = \varepsilon d_1(x),$$

where ε is a small parameter satisfying

$$0 < \varepsilon \ll 1. \qquad (4.16)$$

Based on this RN introduces (formally) the expansions

$$u(x, t) = \varepsilon^2[u_0(x, t) + \varepsilon^2 u_1(x, t) + \varepsilon^4 u_2(x, t) + \dots]$$

$$v(x, t) = \varepsilon[v_0(x, t) + \varepsilon^2 v_1(x, t) + \varepsilon^4 v_2(x, t) + \dots] \qquad (4.17)$$

$$w(x, t) = \varepsilon[w_0(x, t) + \varepsilon^2 w_1(x, t) + \varepsilon^4 w_2(x, t) + \dots]$$

Expanding λ also as

$$\lambda = \varepsilon^2[\lambda_0 + \varepsilon^2\lambda_1 + \varepsilon^4\lambda_2 + \ldots] \qquad (4.18)$$

and substituting in (4.13) we obtain

$$\lambda_0 = u_{0_x} + \frac{1}{2}[v_0^2 + w_0^2]_x \qquad (4.19)$$

Similarly from (4.12)

$$\Lambda = 1 + c_1^2\varepsilon^2\lambda_0 + O(\varepsilon^2).$$

Now RN makes the crucial assumption $c_1^2\varepsilon^2 = \Gamma = O(1)$ by considering the limits

$$\varepsilon \to 0 \quad \text{and} \quad c_1^2 \to \infty \quad \text{but} \quad c_1^2\varepsilon^2 \quad \text{fixed.} \qquad (4.20)$$

For metallic strings ε satisfies (4.16) and $c_1^2 >> 1$. Also for understanding the origin of nonlinear phenomenon, Γ is held fixed in the limit of $\varepsilon \to 0$ and $c_1^2 \to \infty$. Using all the above in (4.15), RN obtains

$$\varepsilon^2(1 - \varepsilon^2 u_{0_x})\frac{\partial u_{0_t}}{\partial t} = \frac{\partial}{\partial x}(1 + \Gamma\lambda_0) + O(\varepsilon^2) \qquad (4.21)$$

$$\varepsilon(1 - \varepsilon^2 u_{0_x})\frac{\partial v_{0_t}}{\partial t} = \varepsilon\frac{\partial}{\partial x}[\{1 + \Gamma\lambda_0\} v_{0_x}] \qquad (4.22)$$

and a similar equation for w. Thus, RN obtains

$$\frac{\partial}{\partial x}\lambda_0 = 0$$

$$\frac{\partial^2 v_0}{\partial t^2} = \frac{\partial^2 v_0}{\partial x^2} + \Gamma\frac{\partial}{\partial x}(v_{0_x}\lambda_0) \qquad (4.23)$$

$$\frac{\partial^2 w_0}{\partial t^2} = \frac{\partial^2 w_0}{\partial x^2} + \Gamma\frac{\partial}{\partial x}(w_{0_x}\lambda_0) .$$

The first equation leads to

$$u_{0_x} = -\frac{1}{2}\left[v_{0_x}^2 + w_{0_x}^2\right] + A_0(t) \qquad (4.24)$$

where $A_0(t)$ is an arbitrary function of time which can be determined from the boundary conditions for u. For the clamped string $u(0) = 0 = u(l)$, thus obtaining

$$u_0 = x\int_0^l \frac{1}{2}\mathbf{v}_{0_x}^2 dx - \int_0^x \frac{1}{2}\mathbf{v}_{0_x}^2 dx.$$

After assuming $\mathbf{f} = \varepsilon' \mathbf{f_0}(\mathbf{x}, \mathbf{t})$ and the damping term to be of the form $\mathbf{g_0} = -2R\mathbf{v}_{0t}$ where R is a damping coefficient, RN obtains the equation (1.1) with

$$\Gamma' = c_1^2 \varepsilon'^2. \tag{4.25}$$

Using

$$\mathbf{f_0}(t) = \sum \mathbf{F}_j(t) sin j\pi x$$

$$\mathbf{v} = \sum \mathbf{V}_j(t) sin j\pi x$$

(1.1) reduces (after ignoring damping) to

$$\ddot{\mathbf{V}}_j + j^2\pi^2\mathbf{V}_j = -\frac{1}{4}\Gamma'j^2\mathbf{V}_j\sum m^2\mathbf{V}_m^2 + \mathbf{F}_j.$$

Truncating it to $j = 1$, RN wrote down the coupled second order nonlinear differential equations for $\mathbf{V}_1(t) = (Y_1(t), Z_1(t))$

$$\ddot{Y}_1 + \pi^2 Y_1 = -\frac{1}{4}\pi^4\Gamma'Y_1(Y_1^2 + Z_1^2) + F_1$$

$$\ddot{Z}_1 + \pi^2 Z_1 = -\frac{1}{4}\pi^4\Gamma'Z_1(Y_1^2 + Z_1^2)$$

and sought a solution in the form $Y_1 = \tilde{Y}_1 cos\omega t$, $Z_1 = \tilde{Z}_1 sin\omega t$, which is possible if

$$(\pi^2 - \omega^2)\tilde{Y}_1 + c_0\Gamma'\tilde{Y}_1^3 = c_1$$

where $c_0 = \frac{3}{16}(1 + \frac{1}{3}\alpha)$ and $c_1 = 1 + \frac{1}{2}\alpha$ with $\alpha = 0$ for planar motion and $\alpha = 1$ for non-planar motion. RN then performs a singular perturbation analysis for the above cubic equation in the limit of $\Gamma' \to 0$ and $\omega \approx \pi$ obtaining critical values for the onset of tubular (non-planar) motion. Next he analyses the stability of planar motion by considering a small non-planar disturbance to an initially planar state of motion, and obtains conditions under which this small disturbance will grow in time. Towards this end he considers the equations for Z_j after including damping

$$\ddot{Z}_j + 2R\dot{Z}_j + \pi^2 Z_j = -\frac{1}{4}\pi^4\Gamma'Z_j\sum m^2Y_m^2$$

and looks for a solution in the form $Y_1 = b_1\varepsilon^{-1}\cos\hat{t}$, $(Y_j = 0 : j \neq 1)$ and $Z_j = e^{-Rt}W_j(t)$ obtaining

$$\ddot{W}_j + (a_j - 2qcos2\hat{t})W_j = 0,$$

where $a_j = \pi^2j^2/\omega^2 - R^2 - 2q$, $q = -\pi^4c_1^2b_1^2/16\omega^2$. An elaborate study is made with a stability chart for Mathieu functions. One of the main conclusions from this study is that non-planar motion results from the instability of planar motion to normal disturbances, due to the naturally occurring space and time variations in tensions in the planar motion.

5. Discussion

We first sketch the circumstances that lead RN to his derivation of (1.1). It all started during RN's visit to his colleague and friend B. S. Ramakrishna's (hereafter BSR) lab at the Electrical and Communications Engineering Department, Indian Institute of Science, Bangalore, where BSR was conducting experiments inspired by the work of Harrison [1948] (Anand 2011, in his obituary says that BSR was a doyen of acoustics with no one else to match his contributions to the field of acoustics in India). BSR pointed out to RN that the linear theory based on the standard linear wave equation predicted infinite amplitude when forced at the natural frequency. However oscillations in the string were never confined to the plane of the forcing at the resonant frequency. Towards this end BSR also derived (cf 2.5) a pair of coupled nonlinear partial differential equations for the two components of the transverse motion but under the assumption that longitudinal displacements were completely ignored. After learning the inconsistent equations obtained by various authors, RN was keen to derive the equations for the motion of strings based on his fluid mechanics background. In addition he felt that the earlier derivations were either not rigorous (the distinguished K. R. Sreenivasan (KRS) who was a student of RN has said that RN brought rigor and class to his research and kindled similar aspirations lying dormant in others) or used assumptions that were unjustified and perhaps even unnecessary; e.g. assuming that the amplitude of the oscillation is small, both tension and amplitudes were small, or that the motion is purely transversal. RN felt that the disagreements as to the correct equations governing the nonlinear behavior of strings was due to different implicit assumptions like, for example, assuming that the longitudinal displacement is everywhere zero. At this point it is worthwhile to point out that many PDE's formulated in the 19th century were derived based on broad physical principles like conservation of mass, momentum and energy, coupled with physical *assumptions* that seemed reasonable but were not rigorously justified, especially in nonlinear problems. Nevertheless it is worthwhile to quote Truesdell [1960] who after reviewing work done till 1788 on elastic bodies says "If I were to draw any summary moral from this development, it would be that mathematics was a more powerful tool than experiment in founding classical theories of elasticity and vibration"

The wave equation was no different and hence there were several attempts to check its validity experimentally. The first one was in 1948 by Harrison and the most recent one by Armstead and Karls in 2006. Also as RN was working in fluid turbulence he wondered about the possibility of what he thought of (at that time) as "mechanical turbulence" (which later came to be subsumed under the word 'chaos' cf O'Reilly and Holmes 1992) in the phenomenon of transfer of energy between different modes by nonlinear coupling as observed by BSR. For instance KRS says "The paper on vibration is a class act in which one can see the stirrings of modern ideas of nonlinear dynamics and chaos −for instance Arnold tongues. RN did not pursue these ideas, and so did not reap the benefit of being a pioneer in those subjects." RN also thought that one could perhaps obtain a turbulence model (many years later RN did formulate one, see Bhat et al 1990) with finite degrees of freedom from the string equation. This was achieved by Miles in 1984, who derived from (1.1) the system

$$\frac{dp_1}{dt} + \alpha p_1 + f(E)q_1 - M(\mathbf{p}, \mathbf{q})p_2 = 0$$

$$\frac{dp_2}{dt} + \alpha p_2 + f(E)q_2 + M(\mathbf{p}, \mathbf{q})p_1 = 0$$

$$\frac{dq_1}{dt} + \alpha q_1 + f(E)p_1 - M(\mathbf{p}, \mathbf{q})q_2 = 1 \qquad (5.26)$$

$$\frac{dq_2}{dt} + \alpha q_2 + f(E)p_2 + M(\mathbf{p}, \mathbf{q})q_1 = 0$$

with

$$\mathbf{p} = (p_1, p_2) \quad \mathbf{q} = (q_1, q_2)$$
$$E = p_1^2 + p_2^2 + q_1^2 + q_2^2$$
$$M(\mathbf{p}, \mathbf{q}) = p_1 q_2 - p_2 q_1$$
$$f(E) = \beta - 3E/2$$

where E and M are measures of energy and angular momentum while α and β are damping and resonance parameters. This system was obtained after taking a single Fourier mode of the form

$$\mathbf{p}(\tau)cos\omega t + \mathbf{q}(\tau)sin\omega t$$

where τ is a slow time "$\varepsilon^{2/3}t$". Miles found that for small enough damping the nonplanar solutions become unstable but he did not find a chaotic attractor. However Bajaj and Johnson [1992] detected numerically the presence of chaotic orbits in the same model. For a complete summary of results for the above system see Molteno and Tufillaro [2004]. Whether these conclusions hold for the original equation (1.1) is not clear as a formal connection between them has not been established. One of the reasons for this late discovery of chaos in vibrating strings is well stated by Tufillaro et al [1995] - "As remarked by most of these authors chaotic motions in strings are not easy to observe. They exist only in a small parameter range and occur not at the forcing frequency, but rather in the slow oscillations of the amplitude envelope. These amplitude modulations are usually only a fraction of the overall transverse displacement. These facts may help to account for the rather late discovery of chaotic vibrations in such a common system".

The first mathematical study of (1.1) was by Bernstein in 1940 who proved local solvability for initial data that was infinitely differentiable and global solvability for analytic functions. Pohožaev [1975] then extended Bernstein's study to several space dimensions and made an abstract formulation of (1.1) reading

$$\ddot{u} + \phi(< Au, u >)Au = 0$$

where $u(t)$ is a Hilbert space valued function with inner product $< ., . >$, and ϕ is a smooth real valued function satisfying

$$\phi(s) \geq \lambda_0 > 0,$$

for a given positive λ_0 and A is a symmetric positive definite operator. Surprisingly neither Bernstein nor Pohožaev quote Kirchhoff. In fact Arosio [1995] says "Bernstein did not quote Kirchhoff's monograph; Bernstein is apparently interested in no physical application...." Even Truesdell (cf Narasimha 1969) does not mention about Kirchhoff when he writes to RN about the misattribution of the total derivative to Lagrange instead of Euler by RN in his JSV paper. As already mentioned above, reference to Kirchhoff's work is more recent and was perhaps inspired by RN. This seems plausible from the work of Dickey [1973], which is the first paper on a mathematical analysis of (1.1) that quotes RN but not Kirchhoff. An earlier paper by Nishida [1971] that studies existence and uniqueness does not refer to Kirchhoff, instead refers to Eisley [1966].

Lions [1978] who refers to RN and Dickey's works quotes Pohožaev but not Kirchhoff or Carrier. In fact, at least till 1985 there is no reference to Kirchhoff's work as Arosio [1985] still calls (1.1) "Carrier-Narasimha equation". It was sometime in the early nineties when more mathematicians started to work on this equation, that the work of Kirchhoff was rediscovered. In the 1993 review Arosio calls (1.1) the Kirchhoff equation and says that Carrier and Narasimha discovered it without knowing Kirchhoff's work. Since then (1990's) there has been much mathematical study of (1.1), nowadays referred to as "Kirchhoff type" equation. For a survey on the mathematical studies of (1.1) we refer to Spagnolo [1992], Arosio [1993], Medeiros et al [2002] and Huet [2010]. Also, that Carrier's equation is not the same as either Kirchhoff or RN's has been realized quite late. Medeiros et al [2002] mention that Cousin et al [1997] generalize the above abstract formulation by considering

$$\ddot{u} + \phi(< A^\theta u, u >)Au = 0$$

for $0 \leq \theta \leq 1$ so that for $\theta = 0$ one gets Carrier's integral term. Nowadays this is more explicitly stated as Carrier's evolution equation, see Limaco and Clark [2007].

We end this section with a brief description of Dickey's work. Although the 1973 work of Dickey referred to RN the equation considered was not (1.1) but instead the local equation

$$\frac{\partial^2 w}{\partial t^2} = (a_0 + a_1 w_x^2)\frac{\partial^2 w}{\partial x^2}.$$

Later however Dickey [1978] considered a special case of (1.1)

$$\frac{\partial^2 w}{\partial t^2} = \left[\lambda + \int_0^\infty w_x^2 dx\right]\frac{\partial^2 w}{\partial x^2}; \quad x > 0$$

$$w(x, 0) = f(x), w_t(x, 0) = g(x), w(0, t) = 0.$$

Taking the one sided Fourier transform $\hat{w}$ of w Dickey obtains the ODE

$$\frac{d^2\hat{w}}{dt^2} = -\xi^2\left[\lambda + \frac{\pi}{2}\int_0^\infty \xi^2 \hat{w}^2 d\xi\right]\hat{w}$$

where Parseval's relation is used for the integral term. Truncating the integral to the finite interval $[0, A]$ and considering

$$\frac{d^2\hat{w}_A}{dt^2} = -\xi^2\left[\lambda + \frac{\pi}{2}\int_0^A \xi^2 \hat{w}_A^2 d\xi\right]\hat{w}_A.$$

Dickey derives bounds for $\hat{w}_A$ which ensure that $\hat{w}_A \to \hat{w}$ as $A \to \infty$ thus establishing existence of a solution to $\hat{w}$, which then establishes existence for w.

6. Conclusions

Based on all the above it is clear that (1.1) is an important equation in describing the nonlinear vibration of strings much more accurately compared not only to the standard wave equation, but also to the other nonlinear string equations proposed in the literature prior to 1968. A similar equation but without considering the damping term, total derivative and lateral displacement was derived by Kirchhoff (1876). RN was unaware (as were many others) of Kirchhoff's work when he derived (1.1) in 1968. It is surprising that even Lord Rayleigh (who lived during the period 1842-1919 and made a theoretical study of resonance in vibrating strings) also seems to be unaware of Kirchhoff's work. Dickey [1973] is the first person to quote RN (but not Kirchhoff) in his mathematical analysis of (1.1). All earlier mathematical studies by Bernstein [1940],

Nishida [1971] and Pohožaev [1975] also do not mention Kirchhoff while referring to the origins of (1.1). This state of un-awareness continued till (at least) 1985. The awakening to Kirchhoff's work seems to have occurred in the 1990's. Although Kirchhoff derived the partial integro-differential equation based on physical arguments, thus depriving himself (unlike RN) of higher order)-proximations, he did make the fundamental point that longitudinal vibrations cannot be neglected. Therefore it would be entirely appropriate to call it the Kirchhoff-Narasimha equation.

It is amazing that RN analysed the vibrations of the string based on simple curiosity regarding the experiments of a colleague, with such profound consequences like leading to a partial integro-differential equation that has been the subject of such intense studies (like existence, uniqueness, regularity and long time behaviour of solutions including the possibility of chaotic behaviour) by mathematicians, engineers and physicists - something unparalleled in the history of mathematical modeling done in India.

7. Acknowledgements

I thank Prof. Roddam Narasimha for sharing generously his thoughts, feelings, ideas and insights.

References

Anand, G. V., 1966, Nonlinear resonances in stretched strings with viscous damping. *J. Acoust. Soc. Am.* **40**, 1517- 1528.

Anand, G. V., 2011, B. S. Ramakrishna (1921-2011). *Current Science.* **100**, 1572-1573.

Antman, S.S, 1980, The equations for large vibration of strings, *Amer. Math. Monthly.* **87**, 359-370.

Armstead, D. C. and Karls, M. A., 2006, Does the wave equation really work?. *PRIMUS* **XVI** 162-177.

Arosio, A., 1985, Global (in time) solution of the approximate nonlinear string equation of G. F. Carrier and R. Narasimha. *Comment. Math. Univ. Carolin.* **26**, 169-172.

Arosio, A., 1995, Averaged evolution equations. The Kirchhoff string and its treatment in scales of Banach spaces. *Functional analytic methods in complex analysis and applications to partial differential equations (Trieste, 1993), 220-254, World Sci. Publ., River Edge, NJ.*

Bajaj, A and Johnson, J., 1992, On the amplitude dynamics and crisis in resonant motion of stretched strings. *Philos. Trans. R. Soc. Lon. Ser. A.* **338**, 1-41.

Bernstein, S., 1940, Sur une classe d'équations fonctionnelles aux dérivées partielles, *Izv. Akad. Nauk SSSR Ser. Mat.* **4**, 17-26.

Bhat, G. S., Narasimha, R. and Wiggins, S., 1990, A simple dynamical system that mimics open-flow turbulence. *Physics of Fluids. A.* **2** 1983-2001.

Bisen, L., 2007, On the Motion of Flexible Strings and Filaments in Inertial and Viscous Regimes, *PhD Thesis. The University of Texas at Austin.*

Carrier, G.F., 1945, On the nonlinear vibration problem of the elastic string. *Q. J. Appl. Math.* **3**, 157-165.

Carrier, G.F., 1949, A note on the vibrating string. *Q. J. Appl. Math.* **7**, 97-101.

Coulson, C. A. and Jefferey, A., 1977 Waves, *Longman, London (First edition: Oliver and Boyd, Edinburgh; Interscience Publishers, Inc., New York, 1941)*

Cousin, A. T., Frota, C. L., Larkin, N. A. and Medeiros, L. A., 1997, On the abstract model of the Kirchhoff-Carrier equation. *Comm. Appl. Anal.* **1**, 389-404.

d'Alembert J., 1747, Recherches sur la courbe que forme une corde tendue mise en vibration. *Mémoire de l'Acadmie Royale des Sciences et Belles-Lettres de Berlin* **3** (**1747/1749**), 214-249.

Dickey, R. W., 1973, The initial value problem for a nonlinear semi-infinite string. *Proc. Amer. Math. Soc.* **37**, 149-156.

Dickey, R. W., 1978, The initial value problem for a nonlinear semi-infinite string. *Proc. Roy. Soc. Edinburgh.* (Sect. **A 82**,) 19-26.

Eisley, J., 1966, Nonlinear deformation of elastic beams, rings and strings. *Applied Mechanics Surveys. Edited by H. Abramson, H. Liebowitz, J. Crowley and S. Juhasz. Spartan Books, Washington D.C.*

Ficken, F. A., 1957, A derivation of the equation for a vibrating string. *The American Mathematical Monthly.* **64**, 155-157.

Harrison, H., 1948, Plane and circular motion of a string. *J. Acoust. Soc. Am.* **20**, 874-875.

Huet, D., 2010, A survey of topics in analysis and partial differential equations. *Int. J. Pure Appl. Math.* **62** 79-128.

Keller, J.B., 1959, Large amplitude motion of a string. *Amer. J. Phys.* **27**, 584-586.

Kevorkian, J., 2000, Partial differential equations. Second Edition. *Texts in Applied Mathematics.* **35**. *Springer Verlag.*

Kirchhoff, G., 1876, Vorlesungen über Mathematische Physik: Mechanik, **Vol. 28.** B. *G. Teubner, Leipzig.*

Kurmyshev, E. V., 2003, Transverse and longitudinal mode coupling in a free vibrating soft string, *Physics Letters (A).* **310**, 148-160.

Leamy, M. J. and Gottlieb, O., 2001, Nonlinear dynamics of a taut string with material nonlinearities. *J. of Vibration and Acoustics,* **123**, pp 53-60.

Lee, E. W., 1957, Nonlinear forced vibration of a stretched string. *British J. of Appl. Physics* **8**, 411-413.

Leissa, A. W. and Saad, A. M., 1994, Large amplitude vibrations of strings. Journal of Applied Mechanics, *Trans. of the ASME,* **61**,296-301.

Limaco, J. and Clark, H. R., 2007, Remarks on Carrier evolution equation. *Mat. Contemp.* **32**, 169-191.

Lions, J. L., 1978, On some equations in boundary value problems of mathematical physics, in: G.M. de La Penha, L.A Medeiros (Eds.), *Contemporary Development in Continuous Mechanics and Partial Differential Equations,* 285-346.

Medeiros, L. A. and Larkin, N. A., 1992, Nonlocal unilateral problem for a nonlinear hyperbolic equation of the theory of elasticity. *Mat. Contemp.* **3**, 109-126.

Medeiros, L. A., Limaco, J. and Menezes, S. B., 2002, Vibrations of elastic strings: Mathematical aspects, Part One. *J. of Comp. Anal. and Appl.* **4**, 91-127.

Miles, J., 1965, Stability of forced oscillations of a vibrating string. *J. Acoust. Soc. Am.* **38**, 855-861.

Miles, J., 1984, Resonant, nonplanar motion of s stretched string. *J. Acoust. Soc. Am.* **75**, 1505-1510.

Molteno, T.C. and Tufillaro, N. B., 2004, An experimental investigation into the dynamics of a string. *Amer. J. Phys.* **72**, 1157-1169.

Murthy, G. S. S. and Ramakrishna, B. S, 1965, Nonlinear character of resonance in stretched strings. *J. Acoust. Soc. Am.* **38**, 461-47.

Narasimha, R., 1968, Nonlinear vibration of an elastic string, *J. Sound and Vibration.* **8**, 134-146.

Narasimha, R., 1969, The equations of motion for a flexible string, *J. Sound and Vibration.* **10**, 350.

Nayfeh, A.H. and Mook, D.T., 1979, Nonlinear Oscillations. *Wiley, New York.*

Nishida, T., 1971, A note on the nonlinear vibrations of the elastic string. *Mem. Fac. Engrg. Kyoto Univ.* **33**, 437-459.

Oplinger, D.W., 1960, Frequency response of a nonlinear stretched string. *J. Acoust. Soc. Am.* **32**, 1529-1538.

O'Reilly, O. and P. J. Holmes 1992, Non-Linear, Non-Planar and Non-Periodic vibrations of a string, *J. Sound and Vibration.* **153**, 413-435.

Osgood, W. F., 1925, Advanced Calculus, *The Macmillian Company, New York.*

Pinsky, M. A., 1998, Partial differential equations and boundary value problems with applications. 3rd ed., *Waveland Press, IL.*

Pohožaev, S. I.,1975, A certain class of quasilinear hyperbolic equations. *Mat. Sb., (N.S.)* **96(138)**, 152-166.

Rayleigh, J. W. S., 1883, On Maintained Vibrations, *Philosphical Magazine.* **XV**, 229-235.

Routh, E. J., 1905, The advanced part of a treatise on the dynamics of a system of rigid bodies. *McMillan and Co.*

Spagnolo, S., 1992, The Cauchy problem for Kirchhoff equations. *Proceedings of the Second International Conference on Partial Differential Equations (Italian) (Milan, 1992). Rend. Sem. Mat. Fis. Milano.* **62**, 17-51.

Sreenivasan, K. R., 2003, Professor Narasimha's contribution to fluid mechanics: a perspective. *Advances in Fluid Mechanics, Proceedings of the symposium held in honor of Prof. Roddam Narasimha's 70th birthday, Editors: Alam, M, Govindarajan, R., Ramesh, O. N. and Sreenivas, K.R.* (http://users.ictp.it/ kıs/pdf/2003 011.pdf).

Stuart, I. M., 1966, String vibrating at finite amplitude. *J. Australian Math. Soc.* **6**, 369-382.

Truesdell, C., 1960, Outline of the history of flexible or elastic bodies to 1788. *J. Acoust. Soc. Am.* **32**, 1647-1656.

Tufillaro, N. B., 1989, Nonlinear and chaotic string vibrations. *American J. of Phys.* **57,** 408-414.

Tufillaro, N. B., Wyckoff, P., Brown, R., Schreiber, T and Molteno, T, 1995, Topological time-series analysis of a string experiment and its synchronized model. *Phys. Rev. E.* **51,** 164-174.

Villaggio, P., 1997, Mathematical models for elastic structures. *Cambridge University Press, Cambridge.*

Watzky, A., 1992, Non-linear three-dimensional large-amplitude damped free vibration of a stiff elastic stretched string. *J. Sound Vibration.* **153,** 125-142.

Wheeler, G.F. and Crummett, W.P., 1987, The Vibrating String Controversy, *Am. J. Phys.* **55,** 33-37.

Yong, D., 2006, Strings, Chains and Ropes, *SIAM Review.* **48,** 771-78.

Representations of Complex Semi-simple Lie Groups and Lie Algebras

Apoorva Khare*

Lie groups and Lie algebras occupy a prominent and central place in mathematics, connecting differential geometry, representation theory, algebraic geometry, number theory, and theoretical physics. In some sense, the heart of (classical) representation theory is in the study of the semisimple Lie groups. Their study is simultaneously simple in its beauty, as well as complex in its richness. From Killing, Cartan, and Weyl, to Dynkin, Harish-Chandra, Bruhat, Kostant, and Serre, many mathematicians in the twentieth century have worked on building up the theory of semisimple Lie algebras and their universal enveloping algebras. Books by Borel, Bourbaki, Bump, Chevalley, Humphreys, Jacobson, Varadarajan, Vogan, and others form the texts for (introductory) graduate courses on the subject.

The purpose of this article is to provide an exposition of the famous 1967 paper [PRV2] by K.R. Parthasarathy, R. Ranga Rao, and V.S. Varadarajan on a class of irreducible Banach space representations of a complex semisimple Lie group. This paper was written in a period containing some of the other classic works in the subject: Harish-Chandra's pioneering work on the principal series representations, and his results on the annihilators of simple modules and central characters; Kostant's work on harmonic polynomials and on his character formula; and papers about Steinberg's formula and Verma modules, to name a few.

In this article, we attempt to explain some of the key ideas and main results of [PRV2]. Given the wide variety of new concepts proposed, as well as its impact on subsequent research in the field, the paper ranks alongside these other works mentioned above.

0.1. The basic motivation for the paper [PRV2] arose out of the important works [Har2, Har3] of Harish-Chandra, in which he constructed a large family of infinite-dimensional irreducible representations of a real semisimple Lie group

*Departments of Mathematics and Statistics, Stanford University, Stanford, CA - 94305, USA. Email: khare@stanford.edu
This work was supported in part by DARPA Grant # YFA N66001-11-1-4131.

G. Harish-Chandra generalized the constructions by Gelfand and Naimark in the case when G is complex semisimple, and his work is regarded today as a cornerstone in the field. For instance, he showed how irreducible representations are subquotients of the principal series representations.

In their work, which followed a few years after [Har2, Har3], Parthasarathy et al returned to the simpler setting of complex semisimple Lie groups, where they were able to use Harish-Chandra's results to obtain a deeper understanding of the structure of Harish-Chandra's Banach space representations of G. Their paper develops a beautiful theory of such representations, each of which decomposes into finite-dimensional modules when restricted to the maximal compact subgroup of G. The authors go on to develop the theory of minimal types, and refine Harish-Chandra's methods (in the complex case) for classifying such irreducible Banach space representations.

0.2. In addition to the above-mentioned primary motivation for [PRV2], the paper develops and proves many other results that have since influenced and inspired a large body of research in the field. We mention a few of these here (and elaborate upon them in future sections). First, the authors provided a multiplicity formula for the classical "tensor product decomposition" problem: given two simple finite-dimensional modules over a complex semisimple Lie algebra, can one write down the decomposition of their tensor product? Combinatorial results due to Kostant, Sternberg, and Brauer were known at the time; however, they required double summations over the Weyl group, computing the Kostant partition function, and cancelling terms in the summation, which made them increasingly harder to implement.

In [PRV2], the authors proposed a formula which was somewhat simpler, directly involving the tensor factors in question. This formula has since been widely used in the literature (as we point out in this article), including in the setting of quantum affine algebras and symmetrizable Kac-Moody algebras, as well as current algebras and other semidirect product Lie algebras.

0.3. Next, a byproduct of this "PRV Theorem" (or formula) was that every such tensor product contains a unique "largest" summand (the "Cartan component"), and a unique "smallest" summand (the "PRV component", or "minimal type"). The former was well-known to be the sum of the two highest weights in question, but the latter was new. Subsequently, the authors and Kostant conjectured the existence of other components, the so-called "generalized PRV components". These are simple modules that occur as direct summands of the tensor product, and their (dominant integral) highest weights are Weyl group-linear combinations of the highest weights of the two tensor factors.

This "PRV Conjecture" has since been proved using multiple techniques in the semisimple as well as Kac-Moody settings. Moreover, it has inspired subsequent research that has led to many contributions in understanding the original problem of computing tensor product multiplicities. Once again, we will discuss these facts in detail below. Throughout this article, we will discuss the results in [PRV2] in the special case of $\mathfrak{sl}_2(\mathbb{C})$, in order to provide a working example - one which we hope will give the reader a greater feel for the results being stated.

0.4. We end this introduction with one last application. In [PRV2], using Kostant's "separation of variables" theorem, the authors defined a set of matrices indexed by pairs of dominant integral weights, whose entries are polynomials on the Cartan subalgebra. The determinants of these matrices yield information about the annihilators of Verma modules, and of their simple quotients. (This is related to the Shapovalov form.)

These "PRV determinants" have since been widely studied, not just in the semisimple case, but in the quantum (affine) and the super-reductive settings as well. In these settings, PRV determinants can be used to determine whether or not the annihilators of Verma modules are generated by their intersection with the center.

1. Notation and Preliminaries

We assume for the purposes of this article that the reader is familiar with basic results concerning the structure of complex semisimple Lie algebras; see [Hu1], for instance. We now set some basic notation, which also serves as a quick summary of the theory. Given a complex semisimple Lie algebra $\mathfrak{g}$, fix a Cartan subalgebra $\mathfrak{h} \subset \mathfrak{g}$ (which is abelian and self-normalizing in $\mathfrak{g}$). Then $\mathfrak{g}$ has a direct sum decomposition: $\mathfrak{g} = \mathfrak{h} \oplus \bigoplus_{\alpha \in R} \mathfrak{g}_\alpha$, where R is the set of roots and $\mathfrak{g}_\alpha$ is the one-dimensional root space for each $\alpha \in R \subset \mathfrak{h}^*$. Here, an element $\lambda \in \mathfrak{h}^*$ is called a weight (in [PRV2] it is called a "rank"), and if M is an $\mathfrak{h}$-module, then its λ-weight space is defined to be:

$$M_\lambda := \{m \in M : h \cdot m = \lambda(h)m \ \forall h \in \mathfrak{h}\}.$$

The weights of M, denoted $\mathrm{wt}(M)$, are those $\lambda \in \mathfrak{h}^*$ for which $M_\lambda \neq 0$. M is a weight module if $M = \bigoplus_{\lambda \in \mathfrak{h}^*} M_\lambda$. For instance, $\mathfrak{g}$ is a $\mathfrak{g}$-module under the adjoint action, and a weight $\mathfrak{h}$-module. The nonzero weights of $\mathfrak{g}$ are precisely the roots: $\mathrm{wt}(\mathfrak{g}) = R \coprod \{0\}$.

A simple example to keep in mind is $\mathfrak{g} = \mathfrak{sl}_2(\mathbb{C})$. This has a basis

$$e = \begin{pmatrix} 0 & 1 \\ 0 & 0 \end{pmatrix}, \qquad f = \begin{pmatrix} 0 & 0 \\ 1 & 0 \end{pmatrix}, \qquad h = \begin{pmatrix} 1 & 0 \\ 0 & -1 \end{pmatrix}$$

of size three, with defining relations:

$$[h, e] = 2e, \qquad [h, f] = -2f, \qquad [e, f] = h.$$

Here, $\mathfrak{h} = \mathbb{C} \cdot h$ and $R = \{\pm\alpha\}$, where $\alpha(h) = 2$. Thus, $\mathfrak{g}_\alpha = \mathbb{C} \cdot e$, $\mathfrak{g}_{-\alpha} = \mathbb{C} \cdot f$, $\mathfrak{g}_0 = \mathbb{C} \cdot h$.

1.1. Weights and lattices.

Let W be the associated Weyl group and $(-, -)$ the Killing form for $\mathfrak{g}$. Then $(-, -)$ induces an isomorphism : $\mathfrak{h} \to \mathfrak{h}^*$. Fix a positive system $R^+ \subset R$ of roots - or equivalently, the subset $\Pi = \{\alpha_i : i \in I\}$ of simple roots indexed by a set I. Then Π is a basis of $\mathfrak{h}^*$, and $R = R^+ \coprod R^-$, where $R^+ = -R^- = R \cap \mathbb{Z}_{\geq 0}\Pi \subset \mathfrak{h}^*$. For each $i \in I$, suppose $h_i' \longleftrightarrow \alpha_i$ via the Killing form; now define the co-roots to be $h_i := (2/\alpha_i(h_i')) \cdot h_i' \in \mathfrak{h}$.

Next, choose Chevalley generators $e_i \in \mathfrak{g}_{\alpha_i}$ and $f_i \in \mathfrak{g}_{-\alpha_i}$ that generate (as above) a copy of $\mathfrak{sl}_2$ together with h_i. Then the e_i and f_i generate nilpotent sub-algebras $\mathfrak{n}^\pm$ of $\mathfrak{g}$, and the corresponding Borel (maximal solvable) subalgebras are: $\mathfrak{b}^\pm := \mathfrak{h} \oplus \mathfrak{n}^\pm$.

We now come to distinguished lattices inside the set of weights. A weight $\lambda \in \mathfrak{h}^*$ is said to be dominant if $\lambda(h_i) \geq 0$ for all $i \in I$, and integral if $\lambda(h_i) \in \mathbb{Z}$. The set of integral weights is a lattice $\Lambda \subset \mathfrak{h}^*$, whose $\mathbb{Z}$-basis is the set of fundamental weights $\{\varpi_i : i \in I\}$. They are defined by: $\varpi_i(h_j) := \delta_{ij}$. Let Λ^+ denote the set of dominant integral weights - which are simply $\mathbb{Z}_{\geq 0}$-linear combinations of the ϖ_i. Then Λ^+ is also (in bijection with) the set of dominant characters of a maximal torus T of G (where G is a complex connected Lie group.such that $\mathfrak{g} = \mathrm{Lie}(G)$).

The weight lattice Λ also contains the root lattice $\mathbb{Z}\Pi$ with $\mathbb{Z}$-basis Π. The group W acts on $\mathfrak{h}^*$ and preserves either lattice. It is generated by the simple reflections $\{s_i : i \in I\}$ which act via: $s_i(\lambda) := \lambda - 2\lambda(h_i)\alpha_i$. The reflections s_i satisfy the Coxeter relations according to the Dynkin diagram of $\mathfrak{g}$, and W is a finite group with a well-defined length-function $\ell : W \to \mathbb{Z}_{\geq 0}$, the associated Bruhat order (see [Hu2, Section 0.4]), and a unique longest element $w_\circ = w_\circ^{-1}$.

For example, for $\mathfrak{g} = \mathfrak{sl}_2(\mathbb{C})$, $\Pi = R^+ = \{\alpha\}$, where $\alpha(h) = 1$. The associated fundamental weight is $\varpi = \frac{1}{2}\alpha$, so that $\mathbb{Z}\Pi = 2\Lambda$ and $\Lambda^+ = \mathbb{Z}_{\geq 0}\varpi$. Moreover, $W = S_2 = \{1, s = w_\circ\}$, where $w_\circ\lambda = -\lambda$ for all weights $\lambda \in \mathfrak{h}^* = \mathbb{C}\varpi = \mathbb{C}\alpha$.

1.2. Finite-dimensional representations.

A representation or module of a group G is simply a group homomorphism $\varpi : G \to GL(V)$ for some (real or complex) vector space V. Similarly, a $\mathfrak{g}$-module V is a Lie algebra homomorphism $\varpi : \mathfrak{g} \to \mathrm{End}(V) = \mathfrak{gl}(V)$. We say that ϖ is irreducible if V has no nonzero proper submodule; completely reducible or semisimple if V is a direct sum of irreducible submodules; and finite-dimensional if $\dim V < \infty$.

When $\mathfrak{g}$ is semisimple, the irreducible finite-dimensional representations are all weight modules for $\mathfrak{h}$, and parametrized by Λ^+. Here is a quick construction: suppose $\mathfrak{Ug}$ is the universal enveloping algebra of $\mathfrak{g}$, and for any $\lambda \in \mathfrak{h}^*$, let I_λ be the left $\mathfrak{Ug}$-ideal generated by $\ker \lambda \subset \mathfrak{h}$ and $\{e_i : i \in I\}$. The Verma module $M(\lambda)$ is defined to be the quotient $\mathfrak{Ug}/I_\lambda$. $M(\lambda)$ is a weight module and $\mathrm{wt}(M(\lambda)) = \lambda - \mathbb{Z}_{\geq 0}\Pi$. Moreover, a cyclic generator of $M(\lambda)$ lies in $M(\lambda)_\lambda = \mathbb{C} \cdot \overline{1_{\mathfrak{Ug}}}$, which is called the "highest weight space".

The modules $M(\lambda)$ were studied by Verma in his thesis and in [Ve]; they are of tremendous importance in representation theory - not only for semisimple Lie algebras, but also Kac-Moody and Virasoro algebras, quantum groups, and other algebras with triangular decomposition. Every Verma module $M(\lambda)$ has a largest maximal submodule and hence a unique simple quotient; denote this by $V(\lambda)$. Then $V(\lambda)$ also has the same properties as $M(\lambda)$ (mentioned in the previous paragraph); moreover, the modules $V(\lambda)$ are pairwise non-isomorphic for $\lambda \in \mathfrak{h}^*$.

Note that $V(\lambda)$ is finite-dimensional if and only if $\lambda \in \Lambda^+$, and these exhaust all finite-dimensional simple $\mathfrak{g}$-modules up to isomorphism. The dual space to a $\mathfrak{g}$-module is also a $\mathfrak{g}$-module; for instance, $V(\lambda)^* \cong V(-w_\circ\lambda)$ if $\lambda \in \Lambda^+$. Moreover, every finite-dimensional $\mathfrak{g}$-module is semisimple; in other words, every indecomposable finite-dimensional $\mathfrak{g}$-module is irreducible.

For example, if $\mathfrak{g} = \mathfrak{sl}_2$, then for every $0 \leq n \in \mathbb{Z}$, there exists a unique irreducible $\mathfrak{sl}_2$-module $V(n) \cong V(n)^*$ of dimension $n + 1$. (Note that we are abusing notation by using $V(n)$ to refer to $V(n\varpi)$.) $V(n)$ contains a vector v_n of weight n (a "highest weight vector"), and a basis $v_{n-2i} := (f^i/i!)v_n$ of weight vectors, for $0 \leq i < \dim V(n)$. One checks that for all i,

$$h \cdot v_{n-2i} := (n-2i)v_{n-2i}, \quad e \cdot v_{n-2i} = (n-i+1)v_{n-2i+2}, \quad f \cdot v_{n-2i} = (i+1)v_{n-2i-2},$$
$$\tag{1.1}$$

where we set $v_{n+2} = v_{-n-2} = 0$. A concrete example of $V(n)$ is provided by the space of homogeneous polynomials in X, Y of total degree n. Define

$$P_n := \ker\left(-n + X\frac{\partial}{\partial X} + Y\frac{\partial}{\partial Y}\right) \subset \mathbb{C}[X, Y].$$

Now define $\rho_n : \mathfrak{sl}_2 \to \mathrm{End}_{\mathbb{C}}(P_n)$ via:

$$\rho_n(e) := X\frac{\partial}{\partial Y}, \qquad \rho_n(f) := Y\frac{\partial}{\partial X}, \qquad \rho_n(h) := X\frac{\partial}{\partial X} - Y\frac{\partial}{\partial Y}.$$

Then $P_n \cong V(n)$ as $\mathfrak{sl}_2$-modules.

1.3. Central characters.

Given an associative algebra A, denote its center by $Z(A)$. An important tool in studying Verma modules and finite-dimensional modules over a (complex) semisimple Lie algebra $\mathfrak{g}$ is the center

$Z(\mathfrak{U}\mathfrak{g})$. Classical results of Chevalley and Harish-Chandra imply that this is a polynomial algebra in $|I|$ (algebraically independent) generators. Moreover, for all $\lambda \in \mathfrak{h}^*$, there exists a central character (i.e., an algebra homomorphism) $\chi(\lambda) : Z(\mathfrak{U}\mathfrak{g}) \to \mathbb{C}$, such that $\ker \chi(\lambda)$ kills $M(\lambda)$ and hence $V(\lambda)$ for all $\lambda \in \mathfrak{h}^*$. In particular, every $z \in Z(\mathfrak{U}\mathfrak{g})$ acts on $V(\lambda)$ by a scalar (for each λ).

The following important results due to Harish-Chandra completely classify and explain better, the set of central characters. (These are also known as *infinitesimal characters* in the literature.) To state the results, we need some notation: define $\rho := \frac{1}{2} \sum_{\alpha \in R^+} \alpha \in \mathfrak{h}^*$; then $\rho = \sum_{i \in I} \varpi_i \in \Lambda^+$ and $w_\circ \rho = -\rho$. Now define the twisted action of W on $\mathfrak{h}^*$ via:

$$w * \lambda := w(\lambda + \rho) - \rho, \qquad \forall w \in W, \lambda \in \mathfrak{h}^*.$$

Then $w * -$ induces an algebra automorphism of $\operatorname{Sym} \mathfrak{h} = P(\mathfrak{h}^*)$ (the space of complex polynomials on $\mathfrak{h}^*$) for all $w \in W$. Moreover, given any $w \in W$, define $\mathfrak{n}_w^\pm := \bigoplus_{\alpha \in R \cap \mathbb{Z}_{\geq 0}(w\Pi)} \mathfrak{g}_{\pm\alpha}$. Then $\mathfrak{g} = \mathfrak{n}_w^- \oplus \mathfrak{h} \oplus \mathfrak{n}_w^+$ for all $w \in W$; for example, when $w = 1$, this decomposition is precisely $\mathfrak{g} = \mathfrak{n}^- \oplus \mathfrak{h} \oplus \mathfrak{n}^+$. Hence $(\mathfrak{U}\mathfrak{g})_0 \subset \mathfrak{U}\mathfrak{h} \oplus \mathfrak{n}_w^-(\mathfrak{U}\mathfrak{g})\mathfrak{n}_w^+$ by the Poincaré-Birkhoff-Witt (PBW) theorem [Hu1]. Define the Harish-Chandra map $\beta^{w\Pi}$ to be the projection : $(\mathfrak{U}\mathfrak{g})_0 \twoheadrightarrow \mathfrak{U}\mathfrak{h} = \operatorname{Sym} \mathfrak{h}$.

Theorem 1.2 ([Har1]). *For all $w \in W$, $\beta^{w\Pi}$ is a ring homomorphism :* $(\mathfrak{U}\mathfrak{g})_0 \twoheadrightarrow \operatorname{Sym} \mathfrak{h}$, *which restricts to a ring isomorphism :* $Z(\mathfrak{U}\mathfrak{g}) \to (\operatorname{Sym} \mathfrak{h})^{(W,*)}$. *Moreover, for all $\lambda \in \mathfrak{h}^*$, $\chi(\lambda) = \lambda \circ \beta^\Pi$. (Here, λ extends to an algebra map on $\operatorname{Sym} \mathfrak{h}$.) Every character of $Z(\mathfrak{U}\mathfrak{g})$ equals $\chi(\lambda)$ for some $\lambda \in \mathfrak{h}^*$. Moreover, $\chi(\lambda) = \chi(\mu) \Leftrightarrow \lambda = w * \mu$ for some $w \in W$.*

For instance, when $\mathfrak{g} = \mathfrak{sl}_2(\mathbb{C})$, $|I| = 1$ and

$$Z(\mathfrak{U}(\mathfrak{sl}_2(\mathbb{C}))) = \mathbb{C}[\Delta], \qquad \Delta = 4fe + h^2 + 2h, \qquad \beta^\Pi(\Delta) = h^2 + 2h.$$

Then for all $z \in \mathbb{C}$, the Casimir element Δ acts on $V(z\varpi)$ as the scalar $\chi(z\varpi)(\Delta) = z^2 + 2z$. Note that $\rho = \varpi$ and $\chi(z\varpi) \equiv \chi(z'\varpi)$ on $Z(\mathfrak{U}\mathfrak{g}) = \mathbb{C}[\Delta]$, if and only if $z + z' = -2$ - i.e., $z'\varpi = s * (z\varpi)$. Similarly, $s * h = -h - 2$, so:

$$\beta^\Pi(\Delta) = (h+1)^2 - 1 = ((s * h) + 1)^2 - 1 = s * \beta^\Pi(\Delta).$$

2. Harish-Chandra Modules

We start our discussion of [PRV2] with the main motivation: the works of Harish-Chandra. The representations studied by Parthasarathy, Ranga Rao, and Varadarajan are known today as (irreducible) *admissible Harish-Chandra modules*. They were first studied in the setting of real semisimple Lie groups by Harish-Chandra in [Har2, Har3].

2.1. For the better part of a century, and since the advent of quantum mechanics, mathematicians have been interested in unitary representations and harmonic analysis of locally compact (abelian) topological groups. One of the basic results in this direction is the Peter-Weyl Theorem, which says that every unitary (Hilbert space) representation of a compact group decomposes as a direct sum of finite-dimensional irreducible submodules. Given the correspondence between complex semisimple groups and compact groups (discovered by Weyl), the class of unitary representations of complex semisimple Lie groups G and their maximal compact subgroups K has been a subject of wide interest and research in the literature.

To explain the motivation for [PRV2], some notation is now needed. Given $G \supset K$ as above, $\mathfrak{k} = \mathrm{Lie}(K)$ is the compact form of $\mathfrak{g} = \mathrm{Lie}(G)$. Let $\mathfrak{g}^{\mathbb{C}} := \mathrm{Lie}(G) \otimes_{\mathbb{R}} \mathbb{C}$ be the complexification of its Lie algebra; this is a complex semisimple Lie algebra that contains the reductive subalgebra $\mathfrak{k}^{\mathbb{C}} := \mathrm{Lie}(K) \otimes_{\mathbb{R}} \mathbb{C}$. In his works cited above, Harish-Chandra initiated the study of a class of irreducible infinite-dimensional G-modules that was larger than the class of unitary G-modules (yet these modules were direct sums of finite-dimensional K-modules). Harish-Chandra constructed and studied these modules algebraically, via their correspondence to $\mathfrak{g}^{\mathbb{C}}$-modules (when G has finite center). This correspondence was known for finite-dimensional modules, but he showed how to extend it to a deep and powerful theory of Banach-space representations of G.

More precisely, given a continuous Banach space G-representation π, whose restriction to K contains every irreducible K-module with at most finite multiplicity (this is called "admissibility"), Harish-Chandra considered its subspace of *K-finite vectors* (i.e., the vectors that lie in finite-dimensional K-stable subspaces) - or more precisely, the $\mathfrak{k}^{\mathbb{C}}$-finite vectors. This subspace is called the *infinitesimal representation associated to* π. Two such Banach space representations are said to be *infinitesimally equivalent* if their infinitesimal representations are equivalent. (For instance, Harish-Chandra showed in [Har2] that two irreducible unitary G-modules are equivalent if and only if they are infinitesimally equivalent.) One of the crown jewels of his work is the *subquotient theorem* [Har3], which says that every such admissible V is infinitesimally equivalent to a subquotient of a Hilbert space representation of G (the "principal series representations").

2.2. We now return to [PRV2], where the authors are interested in using Harish-Chandra's work to gain a deeper understanding of a special case of this situation: namely, when G is already a complex group. (This setting was also studied earlier - by Weyl in relating complex and compact groups, but also by Gelfand and Naimark [GN], and Harish-Chandra as well.) By [Har3], it turns out that every $\mathfrak{k}^{\mathbb{C}}$-finite irreducible G-representation V (with at most finite multiplicities) has an infinitesimal character. In other words, the center

$Z(\mathfrak{U}(\mathfrak{g}^{\mathbb{C}}))$ acts by scalars on it. As noted in [PRV2, Va], the problem of describing the infinitesimal equivalence classes of irreducible Banach space G-representations (which are "admissible", hence equipped with an infinitesimal character) can now be reduced by Harish-Chandra's work, to describing the irreducible $\mathfrak{k}^{\mathbb{C}}$-finite representations - i.e., the so-called simple "$(\mathfrak{g}^{\mathbb{C}}, \mathfrak{k}^{\mathbb{C}})$-modules" (or "$(\mathfrak{g}^{\mathbb{C}}, K)$-modules"). Later in this section, we will mention certain prominent features from Harish-Chandra's approach in the real semisimple case, as we specialize them to the complex case in [PRV2].

Thus, the primary motivation in [PRV2] was to study the irreducible G-representations when G is a complex semisimple group, by applying the methods and deep results from [Har2, Har3]. For instance, the authors are able to simplify Harish-Chandra's description of the closed subspaces of the principal series representations, which yield Banach (actually, Hilbert) space G-representations. Furthermore, the theory of minimal types developed in [PRV2] helps obtain a deeper understanding of these simple $(\mathfrak{g}^{\mathbb{C}}, \mathfrak{k}^{\mathbb{C}})$-modules.

We now introduce the setting of [PRV2]. If G is a complex Lie group, then $\mathfrak{g} := \mathrm{Lie}(G)$ is a complex Lie algebra, and $\mathrm{Lie}(K)$ is its compact (real) form. Thus as real Lie algebras, $\mathfrak{g} = \mathrm{Lie}(K) \oplus \sqrt{-1} \cdot \mathrm{Lie}(K)$ in the complex structure of $\mathfrak{g}$. The complexified pair $(\mathfrak{g}^{\mathbb{C}}, \mathfrak{k}^{\mathbb{C}})$ is isomorphic to $(\mathfrak{g} \times \mathfrak{g}, \overline{\mathfrak{g}})$, where $\overline{\mathfrak{g}}$ is the diagonal copy of $\mathfrak{g}$ embedded in $\mathfrak{g} \times \mathfrak{g}$.[1] Now the $\mathfrak{g} \times \mathfrak{g}$-modules of interest (studied by Harish-Chandra in general) are the ones that decompose into direct sums of finite-dimensional $\overline{\mathfrak{g}}$-modules with at most finite multiplicities.

Here is the precise framework studied in [PRV2] (and henceforth in this article), stated here in a slightly more general setting.

Definition 2.1. Suppose $\mathfrak{g}$ is a complex reductive finite-dimensional Lie algebra contained in a complex Lie algebra $\widehat{\mathfrak{g}}$. Define the category $\mathcal{C}(\widehat{\mathfrak{g}}, \mathfrak{g})$ to be the full subcategory of $\widehat{\mathfrak{g}}$-modules, such that every object is isomorphic to a direct sum of finite-dimensional irreducible $\mathfrak{g}$-modules $\mathcal{D}$, each of which occurs with finite multiplicity. (This last condition is termed $\mathfrak{g}$-*admissibility*.) Define $[V : \mathcal{D}]$ to be this multiplicity (which may be zero if no summand is isomorphic to $\mathcal{D}$); this integer does not depend on the direct sum decomposition of V. (Note that we assume $\mathcal{D} \neq 0$.)

If V is in $\mathcal{C}(\widehat{\mathfrak{g}}, \mathfrak{g})$ and $\mathcal{D}$ is a (nonzero) simple $\mathfrak{g}$-module, the *isotypical subspace* $V_{\mathcal{D}}$ of V is defined as the (finite-dimensional) span of all the $\mathfrak{g}$-submodules of V that are isomorphic to $\mathcal{D}$. Clearly, the center $Z(\mathfrak{U}\widehat{\mathfrak{g}})$ preserves $V_{\mathcal{D}}$ for each $\mathcal{D}$, and hence acts locally finitely on V. Moreover, $[V : \mathcal{D}]$ then equals $[V_{\mathcal{D}} : \mathcal{D}] = \dim V_{\mathcal{D}} / \dim \mathcal{D}$.

---

[1] This is achieved using a conjugation $X \mapsto X^c$ of $\mathfrak{g}$ that fixes $\mathfrak{k}$. Thus, $\mathfrak{g}$ embeds inside $\mathfrak{g}^{\mathbb{C}}$ via: $X \mapsto (X^c, X)$, and when restricted to $\mathfrak{k}$, one obtains: $X \mapsto (X, X)$ - whence we get that $\mathfrak{k}^{\mathbb{C}} = \overline{\mathfrak{g}}$.

2.3. Examples of Harish-Chandra modules in the literature.

The goal of [PRV2] was to study the simple objects in the category $\mathcal{C}(\mathfrak{g} \times \mathfrak{g}, \bar{\mathfrak{g}})$. Before elaborating on their results, we remark that various families of Harish-Chandra modules have been widely studied in the literature. For example, Harish-Chandra modules are examples of integrable $\bar{\mathfrak{g}}$-modules - i.e., $\bar{\mathfrak{g}}$-modules where every vector is contained in a finite-dimensional $\bar{\mathfrak{g}}$-module. Here are some other examples; in them, we always assume that $\mathfrak{g}$ is semisimple (and complex).

Example 2.2. Suppose $\mathfrak{g}$ is semisimple and $\mathfrak{g}_0$ is its compact real form. Let G_0 be a compact Lie group with Haar measure μ_{G_0}, such that $\mathfrak{g}_0 = \mathrm{Lie}(G_0)$. Then by the Peter-Weyl Theorem, a dense subspace V of $L^2(G_0, \mathbb{C}, \mu_{G_0})$ is an object of $\mathcal{C}(\mathfrak{g}, \mathfrak{g})$. Moreover, $[V : V(\lambda)] = \dim V(\lambda)$ for all $\lambda \in \Lambda^+$.

Example 2.3. Recall that $\mathfrak{Ug}$ is a direct sum of finite-dimensional $\mathfrak{g}$-modules, since every term in its standard filtration is. Is it also an object of $\mathcal{C}(\mathfrak{g}, \mathfrak{g})$? The answer is no - in fact, *no* finite-dimensional module occurs with finite nonzero multiplicity. To see this, note by Kostant's "separation of variables theorem" [Ko2] that $\mathfrak{Ug}$ is free as a module (under multiplication) over its center:

$$\mathfrak{Ug} \cong \mathbb{H}(\mathfrak{g}) \otimes Z(\mathfrak{Ug}), \qquad (2.4)$$

where $\mathbb{H}(\mathfrak{g})$ is (isomorphic as a $\mathfrak{g}$-module to) the space of "harmonic polynomials on $\mathfrak{g}$", and is stable under the adjoint action of $\mathfrak{g}$ on $\mathfrak{Ug}$. Thus, the multiplicity in $\mathfrak{Ug}$ of every finite-dimensional module is either 0 or $\dim Z(\mathfrak{Ug})$, which is infinite. In particular, $\mathfrak{Ug}$ is not admissible.

However, $\mathbb{H}(\mathfrak{g})$ is indeed an object in $\mathcal{C}(\mathfrak{g}, \mathfrak{g})$; in fact, Kostant proved in [Ko2] that $[\mathbb{H}(\mathfrak{g}) : V(\lambda)] = \dim V(\lambda)_0$ for all $\lambda \in \Lambda^+$. This is the starting point for another important contribution of [PRV2] to the theory of semisimple and affine (quantized) Lie algebras - the so-called "PRV determinants". We will discuss these in Section 5.

Example 2.5. Simple finite-dimensional $\mathfrak{g} \times \mathfrak{g}$-modules are clearly in $\mathcal{C}(\mathfrak{g} \times \mathfrak{g}, \bar{\mathfrak{g}})$, by Weyl's Theorem of complete reducibility. This example is also the starting point for a result and a conjecture from [PRV2] (the "PRV Theorem" and "PRV conjecture"), that have since been extensively used and generalized in the literature. We address these in detail in the next section.

Example 2.6. The above example of ρ_n for $\mathfrak{g} = \mathfrak{sl}_2(\mathbb{C})$ can be used to produce an object in $\mathcal{C}(\mathfrak{sl}_2(\mathbb{C}), \mathfrak{sl}_2(\mathbb{C}))$ as follows: the $\mathfrak{sl}_2(\mathbb{C})$-module

$$\mathbb{C}[X, Y] = \bigoplus_{n \geq 0} P_n = \bigoplus_{n \geq 0} V(n)$$

is clearly such an object.

Note that $P_n = V(n) = \mathrm{Sym}^{n-1}(V(1))$ for all $n \in \mathbb{N}$. Thus, $\mathbb{C}[X,Y] = \mathrm{Sym}(V(1))$. With this in mind, we can generalize the above example to $\mathfrak{g} = \mathfrak{sl}_n(\mathbb{C})$, as it acts on its simple module $\mathbb{C}^n$ (for $n > 1$). Consider the modules $\mathrm{Sym}^k(\mathbb{C}^n) \subset (\mathbb{C}^n)^{\otimes k}$ for $k \geq 0$. Identifying a basis of $\mathbb{C}^n$ with commuting variables $X_1, \ldots, X_n$, it is not hard to show that as $\mathfrak{g}$-modules, $\mathrm{Sym}^k(\mathbb{C}^n)$ is precisely the space $P_{n,k}$ of homogeneous polynomials in $X_1, \ldots, X_n$ of total degree k, where e_{ij} acts on $P_{n,k}$ as $X_i \partial_j$ for all $1 \leq i, j \leq n$ and all k.

One can now check that $P_{n,k}$ is a simple module over $\mathfrak{sl}_n(\mathbb{C})^2$. Since $n > 1$, hence $\dim P_{n,k} = \binom{k+n-1}{n-1}$ is increasing in k. Thus the $P_{n,k}$ are non-isomorphic for fixed n, and so

$$\mathbb{C}[X_1, \ldots, X_n] = \bigoplus_{k \geq 0} P_{n,k} = \mathrm{Sym}(\mathbb{C}^n)$$

is indeed an object in $\mathcal{C}(\mathfrak{sl}_n(\mathbb{C}), \mathfrak{sl}_n(\mathbb{C}))$.

Example 2.7. If $\mathfrak{g}$ is semisimple and $\mathfrak{h}$ is the Cartan subalgebra of $\mathfrak{g}$, then $\mathcal{C}(\mathfrak{g}, \mathfrak{h})$ is the category of (admissible) weight modules. There has been extensive research on the study and classification of irreducible (admissible) weight modules; see [Ma3] for more on this. We remark that Mathieu also studied Harish-Chandra modules in other settings in [Ma2]: the Virasoro algebra, the Cartan algebra, and the affine Kac-Moody algebras (as mentioned in the conclusion to *loc. cit.*).

Moreover, a very special family of admissible weight modules constitutes the objects of the Bernstein-Gelfand-Gelfand Category $\mathcal{O}$, which was introduced in [BGG]. A lot of research has been undertaken on the Category $\mathcal{O}$ in various settings in modern representation theory - including semisimple and Kac-Moody Lie algebras, the quantum groups associated with them, the Virasoro algebra, and more modern constructions such as rational Cherednik algebras, infinitesimal Hecke algebras, and W-algebras. In particular, for semisimple $\mathfrak{g}$, the classification of irreducible admissible weight modules (by work of Mathieu [Ma3] and others) as well as of primitive ideals (by work of Duflo [Du3] and others) reduces to the study of simple objects in $\mathcal{O}$. See [Hu2, Jo2, Kh, MP] for additional references and results.

2.4. A key class of homomorphisms.

We now outline Harish-Chandra's strategy for studying admissible irreducible G-representations, as it is used by Parthasarathy et al in the complex setting. Given $\mathfrak{g} \subset \widehat{\mathfrak{g}}$ as above, let Ω denote the centralizer of $\mathfrak{g}$ in $\mathfrak{U}\widehat{\mathfrak{g}}$. (This is denoted by $\mathfrak{D}$ in [Va].) Then

---

[2]See http://math.stackexchange.com/questions/120338 for the sketch of a proof.

$Z(\mathfrak{U}\mathfrak{g}) + Z(\mathfrak{U}\widehat{\mathfrak{g}}) \subset \Omega \subset \mathfrak{U}\widehat{\mathfrak{g}}$ is a chain of algebras[3]. Now suppose V is an object of $\mathcal{C}(\widehat{\mathfrak{g}}, \mathfrak{g})$, and $\mathcal{D}$ is a simple finite-dimensional $\mathfrak{g}$-module such that $[V : \mathcal{D}] = r > 0$. Then $\mathcal{D} \cong V(\lambda)$ as finite-dimensional (and hence, highest-weight) $\mathfrak{g}$-modules, and $V_{\mathcal{D}} \cong V(\lambda) \otimes \mathbb{C}^r$ as a $\mathfrak{g}$-module (i.e., $\mathbb{C}^r$ is the multiplicity space). Multiplication by every $z \in \Omega$ preserves the highest weight space $(V_{\mathcal{D}})_\lambda = V(\lambda)_\lambda \otimes \mathbb{C}^r$; this yields a representation $\eta_{V,\mathcal{D}}$ of Ω into $\mathbb{C}^r$. (This is called $\eta((\nu^0), \pi)$ in [PRV2], where $\pi = V$ and $(\nu^0) = \mathcal{D}$.) Moreover $V_{\mathcal{D}}$ now decomposes as $\mathcal{D} \otimes \mathbb{C}^r$, under the joint action of $\mathfrak{g}$ and Ω.

As a special case, suppose $r = 1$. Then $\eta_{V,\mathcal{D}}$ is a homomorphism : $\Omega \to \mathbb{C}$. These homomorphisms are the key tools used in [PRV2] to study simple admissible Harish-Chandra modules, as we now explain.

Suppose $\widehat{\mathfrak{g}} = \mathfrak{g} \times \mathfrak{g} \supset \overline{\mathfrak{g}}$. In order to study simple objects in $\mathcal{C}(\mathfrak{g} \times \mathfrak{g}, \overline{\mathfrak{g}})$, the authors of [PRV2] follow the approach suggested by Harish-Chandra in [Har3]: if $r = [V : \mathcal{D}] > 0$, then as above, $V_{\mathcal{D}} \cong \mathcal{D} \otimes \mathbb{C}^r$ under the joint action of $\mathfrak{U}\overline{\mathfrak{g}}$ and Ω - and moreover, the Ω-representation $\eta_{V,\mathcal{D}}$ is simple. Now the following remarkable fact holds: *the equivalence class of the representation $\eta_{V,\mathcal{D}}$ of Ω determines that of V, for every component $\mathcal{D}$ with $r > 0$*. More precisely, if V, V' are simple objects of $\mathcal{C}(\mathfrak{g} \times \mathfrak{g}, \overline{\mathfrak{g}})$ and $\mathcal{D}$ is a simple finite-dimensional $\overline{\mathfrak{g}}$-module such that $[V : \mathcal{D}] + [V' : \mathcal{D}] > 0$, then

$$[V : \mathcal{D}] = [V' : \mathcal{D}] > 0, \ \eta_{V,\mathcal{D}} \cong_\Omega \eta_{V',\mathcal{D}} \Longleftrightarrow V \cong V'. \tag{2.8}$$

(See [LMC] for a generalization of this fact.) Thus, a "first approach" would be to fix various $\mathcal{D}$ and study the Ω-modules $\eta_{V,\mathcal{D}}$ for all V with $[V : \mathcal{D}] > 0$. The authors remark in [PRV2] that such an approach was not very fruitful and so a different method had to be adopted. Their contribution was to introduce and study the following notion.

Definition 2.9. Suppose V is a simple object in $\mathcal{C}(\mathfrak{g} \times \mathfrak{g}, \overline{\mathfrak{g}})$ and $\lambda \in \Lambda^+$. We say that λ (or $V_{\overline{\mathfrak{g}}}(\lambda)$) is a *minimal type* of V if $[V : V_{\overline{\mathfrak{g}}}(\lambda)] > 0$, and

$$[V : V_{\overline{\mathfrak{g}}}(\mu)] > 0 \implies \lambda \in \text{wt} \, V_{\overline{\mathfrak{g}}}(\mu).$$

It is clear that there is at most one minimal type for each V.

Now the strategy is as follows: first study a class of modules V for which the minimal type $\mathcal{D}$ can be shown to exist. These are the irreducible finite-dimensional representations of $\mathfrak{g} \times \mathfrak{g}$, and it turns out that there is an explicit recipe to compute the homomorphism $\eta_{V,\mathcal{D}}$ in this case. This recipe involves $\eta_{V,\mathcal{D}}$ equalling a polynomial-valued homomorphism $\mathbf{h}^{\Pi'} : \Omega \to P(\mathfrak{h}^* \times \mathfrak{h}^*)$,

---

[3]In Varadarajan's reminiscences [Va], he points out on Page (xii) that Ω is highly non-abelian, so that the first inclusion is not an equality in general.

evaluated at some $\lambda \in \Lambda^+, \nu \in \Lambda$ - in other words, $\mathbf{h}^{\Pi'}(-; \lambda, \nu) : \Omega \to \mathbb{C}$. (This is explained in Section 4.4 and beyond.)

The authors then go on to explicitly construct a family $\{\widehat{\pi}_{\lambda,\nu} : \nu \in \Lambda, \lambda \in \mathfrak{h}^*\}$ of simple objects in $\mathcal{C}(\mathfrak{g} \times \mathfrak{g}, \overline{\mathfrak{g}})$, each of which has a minimal type $\overline{\nu}$ occurring with multiplicity $r = 1$. (This family necessarily includes the finite-dimensional simple $\mathfrak{g} \times \mathfrak{g}$-modules, as we will see below.) The authors show that for each such λ and ν, the related "key homomorphism" $\eta_{\widehat{\pi}_{\lambda,\nu}, \overline{\nu}} : \Omega \to \mathbb{C}$ turns out to be the same recipe $\mathbf{h}^{\Pi'}$ as above, now evaluated at (more general points) λ, ν. Thus, Equation (2.8) can be applied to discuss the classification of these modules $\widehat{\pi}_{\lambda,\nu}$. (Here and henceforth, we abuse notation and use $\eta_{\widehat{\pi}_{\lambda,\nu}, \overline{\nu}}$ to refer to $\eta_{\widehat{\pi}_{\lambda,\nu}, V_{\overline{\mathfrak{g}}}(\overline{\nu})}$.)

Thus, the starting point for [PRV2] - and even earlier, for Varadarajan and Varadhan in 1963 (see [Va]) for the special case of $\mathfrak{g} = \mathfrak{sl}_n(\mathbb{C})$ - was to prove the assertion that finite-dimensional irreducible $\mathfrak{g} \times \mathfrak{g}$-modules have minimal types. Note that such a simple module has highest weight in $\Lambda^+ \times \Lambda^+ \subset (\mathfrak{h} \times \mathfrak{h})^*$, so we can write it as $V(\lambda, \mu)$. It is clear that for all $X \in \mathfrak{g}$, its image in

$$\overline{\mathfrak{g}} \subset \mathfrak{g} \times \mathfrak{g} \subset \mathfrak{U}(\mathfrak{g} \times \mathfrak{g}) = \mathfrak{U}\mathfrak{g} \otimes \mathfrak{U}\mathfrak{g}$$

is precisely $X \otimes 1 + 1 \otimes X$. Thus, restricting $V(\lambda, \mu)$ to $\overline{\mathfrak{g}}$ amounts to considering the module $V_{\overline{\mathfrak{g}}}(\lambda) \otimes V_{\overline{\mathfrak{g}}}(\mu)$. In other words, the study of the minimal type in this setting involves computing the direct summands of the tensor product - i.e., computing Clebsch-Gordan coefficients. This classical problem is the focus of the next section.

2.5. Digression on minimal type due to Vogan.
We end this section with a few remarks on the notion of minimal type. The more widely accepted notion of *minimal K-type* (or *lowest K-type*) in the literature is due to Vogan [Vo], and differs from the above notion (in [PRV2]). More precisely, Vogan defines a weight $\lambda \in \Lambda^+$ to be a *lowest K-type* for an admissible Harish-Chandra (G, K)-module M, if:

1. $V(\lambda)$ is a K-submodule of M; and

2. Among all μ such that $V(\mu)$ is a K-submodule of M, the scalar $(\mu + 2\rho, \mu + 2\rho)$ is minimized at $\mu = \lambda$.

One can now ask what is the relation between these two notions. Note that the definition due to Vogan guarantees existence of the minimal type (for irreducible admissible representations), but not uniqueness. In fact, uniqueness does not hold when G is a (general) real group, such as $SL(2, \mathbb{R})$. However, uniqueness of the minimal K-type is guaranteed if G is a complex group; see [Zh2] for more details.

On the other hand, the definition in [PRV2] guarantees uniqueness but not existence. If this version of the minimal type does exist, then it is necessarily a minimal type due to Vogan. This can be shown using the following generalization of [Hu1, Lemma 13.4.C] (whose proof is the same as that of *loc. cit.*), with $A = \operatorname{wt} V(\mu)$ for any finite-dimensional K-submodule $V(\mu)$ of M:

Proposition 2.10. *Suppose $A \subset \Lambda$ is W-stable, with highest weight μ. (In other words, $A \subset \mu - \mathbb{Z}_{\geq 0}\Pi$.) Fix $\lambda \in A$ and $0 < c \in \mathbb{R}$. Then $\mu \in \Lambda^+$ and A is finite; moreover,*

$$(\lambda + c\rho, \lambda + c\rho) \leq (\mu + c\rho, \mu + c\rho),$$

with equality if and only if $\lambda = \mu$.

3. Tensor Products, Minimal Types, and the (K)PRV Conjecture

In this section, we discuss [PRV2, Section 2.2], which contains several results, as well as a related conjecture, that have been extremely influential on subsequent research in the field. These results and the conjecture have to do with the classical question of computing tensor product multiplicities. Although the primary motivation of Parthasarathy et al was to study tensor products in order to prove the existence and uniqueness of minimal types, some of these statements have been subsequently generalized and have contributed to several aspects of the multiplicity problem. We will list some of the relevant papers and results presently.

3.1. Tensor product multiplicities and the PRV Theorem.

We start by recalling the notion of *Littlewood-Richardson coefficients*. By Weyl's theorem of complete reducibility, given $\lambda, \mu \in \Lambda^+$, we can decompose

$$V(\lambda) \otimes V(\mu) = \bigoplus_{\nu \in \Lambda^+} m_{\lambda,\mu}^{\nu} V(\nu),$$

where the multiplicities $m_{\lambda,\mu}^{\nu} = m_{\mu,\lambda}^{\nu}$ are the coefficients in question, also known as tensor product multiplicities. (In the rest of this article, we will abuse notation and denote $V_{\bar{\mathfrak{g}}}(\lambda)$ by $V(\lambda)$.) If $m_{\lambda,\mu}^{\nu} > 0$, we say that $V(\nu)$ is a *component* of $V(\lambda) \otimes V(\mu)$. For example, $V(\lambda + \mu)$ is always a component, generated by $V(\lambda)_{\lambda} \otimes V(\mu)_{\mu}$, and $m_{\lambda,\mu}^{\lambda+\mu} = 1$ for all $\lambda, \mu \in \Lambda^+$.

The determination of the multiplicities $m_{\lambda,\mu}^{\nu}$ is a longstanding open problem in the literature - as is the simpler problem of computing whether or not $m_{\lambda,\mu}^{\nu}$ is positive. Efforts to answer these questions have been ongoing since even

97

before [PRV2]. For instance, in his famous paper [Ko1], Kostant proved his multiplicity formula, and also showed a necessary condition for $V(\nu)$ to be a component: it must be of the form $\nu = \lambda + \mu_1 \in \Lambda^+$ for some $\mu_1 \in \mathrm{wt}(V(\mu))$. Moreover, $m_{\lambda,\mu}^{\nu} \leq \dim V(\mu)_{\nu-\lambda}$. By work [St] of Steinberg (using Kostant's multiplicity formula), the following was also known:

$$m_{\lambda,\mu}^{\nu} = \sum_{w \in W} \mathrm{sn}(w) \dim V(\mu)_{w*\nu-\lambda} = \sum_{w,w' \in W} \mathrm{sn}(w)\,\mathrm{sn}(w')\mathcal{P}(w'(\mu+\rho) - w(\nu+\rho) + \lambda).$$

Here, $\mathcal{P} : \Lambda^+ \to \mathbb{N}$ is the Kostant partition function (which is also defined to be zero on $\Lambda \setminus \Lambda^+$), and sn $: W \to \{\pm 1\}$ is the sign homomorphism, which is -1 on all simple reflections s_i. Steinberg's results imply [Ku5] that if $(\lambda + \mu')(h_i) \geq -1$ for all $\mu' \in \mathrm{wt}(V(\mu))$ and $i \in I$, then $m_{\lambda,\mu}^{\nu} = \dim V(\mu)_{\nu-\lambda}$. Kostant had shown a special case of this result in [Ko1], where he assumed that $(\lambda + \mu')(h_i) \geq 0$.

In [PRV2], the following multiplicity formula is proved. Given $\mu, \nu \in \Lambda^+$ and $\gamma \in \mathfrak{h}^*$, define:

$$
\begin{aligned}
V^+(\mu; \gamma, \nu) &:= \{v \in V(\mu)_\gamma : e_i^{\nu(h_i)+1} v = 0 \;\forall i \in I\}, \\
V^-(\mu; \gamma, \nu) &:= \{v \in V(\mu)_\gamma : f_i^{\nu(h_i)+1} v = 0 \;\forall i \in I\}.
\end{aligned}
$$

Theorem 3.1 ([PRV2]). *For all $\lambda, \mu, \nu \in \Lambda^+$,*

$$m_{\lambda,\mu}^{\nu} = \dim V^+(\mu; \nu - \lambda, \lambda) = \dim V^+(\nu; \lambda + w_\circ\mu, -w_\circ\mu).$$

Now given $\gamma \in \mathfrak{h}^*$, $\dim V^+(\mu; \gamma, \nu) = \dim V^-(\mu; w_\circ\gamma, -w_\circ\nu)$.

Here is a typical application of this result, which shows how to compute multiplicities.

Example 3.2. Suppose $\lambda, \nu \in \Lambda^+$. If $[V(\lambda) \otimes V(\lambda)^* : V(\nu)] > 0$, then $\lambda - w_\circ\lambda - \nu \in \mathbb{Z}_{\geq 0}\Pi$ by Kostant's results, which implies that $\nu \in \Lambda^+ \cap \mathbb{Z}\Pi$. For every such ν, Theorem 3.1 now says:

$$m_{\lambda,-w_\circ\lambda}^{\nu} = \dim V^+(\nu; 0, \lambda) := \dim\{v \in V(\nu)_0 : e_i^{\lambda(h_i)+1} v = 0 \;\forall i \in I\}.$$

Thus if $\lambda(h_i)$ is large enough for all i (e.g., $\lambda = n\rho$ for $n \gg 0$), then

$$[V(\lambda) \otimes V(\lambda)^* : V(\nu)] = \dim V(\nu)_0 > 0, \qquad \forall \nu \in \Lambda^+ \cap \mathbb{Z}\Pi.$$

where the last inequality follows by a result from [Hal], used below to prove Proposition 3.6. $\qquad\square$

The advantage of the "PRV Theorem" 3.1 over some of the earlier formulae in the literature is that it calculates the multiplicities directly and without cancellation. For instance, note that the above result of Steinberg involves a double summation over the Weyl group - and cancellations of terms - and hence is not suitable for practical computations. Several years prior to [PRV2], Brauer had proposed another such formula in [Br]; it is similar to a result of Klimyk in [Kl], which appeared in the same year as [PRV2]. The result is stated as Exercise 24.9 in [Hu1][4]. It computes the multiplicities $m^\nu_{\lambda,\mu}$ as sums of dimensions $\dim V(\mu)_{\mu'}$, but with coefficients of ± 1 and 0, which again implies the need to perform cancellation (of formal characters).

It is mentioned in [PRV2] that Kostant had obtained Theorem 3.1 previously but had not published it; for a historical account of this result, see [Ko4].

The PRV Theorem 3.1 has been widely used and generalized in the literature. In [CP], Chari and Pressley extend a special case of this result to show that simple integrable modules over affine Lie algebras are quotients of tensor products. In [Jo2], Joseph studies a similar result for a general symmetrizable (quantum) Kac-Moody Lie algebra - as does Mathieu in [Ma1]. Among other applications, Young and Zegers start from Theorem 3.1 in [YZ] and relate Dorey's rule to q-characters of fundamental representations of quantum affine algebras of type ADE. Panyushev and Yakimova study variants and consequences of the result in [PY].

From a personal viewpoint, the author has used the PRV Theorem in his paper [CKR] with Chari and Ridenour, to provide examples of families of finite and infinite-dimensional Koszul algebras which naturally arise out of module categories over semidirect products $\mathfrak{g} \ltimes V(\lambda)$. The result was also used by Chari and Greenstein [CG1, CG2] in the study of representations of the truncated current algebra $\mathfrak{g}[t]/(t^2)$, as well as in other works of Chari and her collaborators, and of Greenstein. These papers have applications in the study of Kirillov-Reshetikhin modules over quantum affine algebras.

3.2. Minimal type. We again start by considering the decomposition of the tensor product into its simple module components. Consider the example where $\mathfrak{g} = \mathfrak{sl}_2(\mathbb{C})$ and $0 \leq \mu \leq \lambda \in \mathbb{Z}_{\geq 0}$. (Once again, we abuse notation and use $\lambda \in \mathbb{Z}$ to refer to $\lambda \cdot \varpi \in \Lambda$.) Recall the well-known *Clebsch-Gordan formula* for $\mathfrak{sl}_2(\mathbb{C})$:

$$V(\lambda) \otimes V(\mu) = V(\lambda + \mu) \oplus V(\lambda + \mu - 2) \oplus \cdots \oplus V(\lambda - \mu). \quad (3.3)$$

We see that there are two distinguished components in this direct sum:

- The "largest" component is $V(\lambda + \mu)$, in that every highest weight occurring on the right, belongs to $\operatorname{wt} V(\lambda + \mu)$. This is called the *Cartan*

---

[4]See also: http://mathoverflow.net/questions/85593/

component or the *maximal type*, and occurs with multiplicity 1. It is generated by the one-dimensional vector space $V(\lambda)_\lambda \otimes V(\mu)_\mu$, which is the tensor product of the two highest weight spaces.

- The "smallest" component is $V(\lambda - \mu)$, in that $\lambda - \mu$ is a weight of every summand occurring on the right. This is called the *PRV component* (after the authors of [PRV2]) or the *minimal type*, and it also occurs with multiplicity 1.

It is reasonable to ask if these results extend to all semisimple $\mathfrak{g}$. Remarkably, the authors of [PRV2] found the answer of this question to be positive! To understand it, one must first make sense of what the minimal type is for general $\mathfrak{g}$. Note above that we could have interchanged λ and μ, since the tensor product is "commutative" (i.e., the Hopf algebra $\mathfrak{Ug}$ is cocommutative). Thus, to choose the minimal type, one chooses the dominant integral weight from among $\{\lambda - \mu, \mu - \lambda\} = W(\lambda + w_\circ\mu)$. Supporting evidence is obtained from Theorem 3.1, where if $\lambda + w_\circ\mu \in \Lambda^+$, then substituting it for ν yields:

$$m_{\lambda,\mu}^{\lambda+w_\circ\mu} = \dim V^+(\lambda + w_\circ\mu; \lambda + w_\circ\mu, -w_\circ\mu) = 1. \tag{3.4}$$

This led Varadarajan and Varadhan to generalize the existence of the minimal type to $\mathfrak{g} = \mathfrak{sl}_n(\mathbb{C})$ for all n, while they were at the Indian Statistical Institute, Kolkata. (See [Va] for a very nice historical account of the development of [PRV2].) Subsequently in [PRV2], the authors extended the result to arbitrary semisimple $\mathfrak{g}$, and obtained the previously sought-for existence and unique multiplicity of the minimal type. Here is their result.

Theorem 3.5 ([PRV2]). *Suppose $\mathfrak{g}$ is semisimple, and $\lambda, \mu \in \Lambda^+$. Given $\nu \in \Lambda^+$, define $\bar{\nu}$ to be the unique W-translate of ν that lies in Λ^+. Then $m_{\lambda,\mu}^{\overline{\lambda+w_\circ\mu}} = 1$. Moreover,*

$$m_{\lambda,\mu}^\nu > 0 \implies \overline{\lambda + w_\circ\mu} \in \mathrm{wt}\, V(\nu).$$

(More generally - say by the result from [KLV] stated in the next proof below - wt $V(\overline{\lambda + w_\circ\mu}) \subset$ wt $V(\nu)$ for all such ν.) Thus, the sought-for minimal type exists and possesses the desired properties. We will see later, how this leads to the construction of interesting polynomial maps and infinite-dimensional Banach space representations of G.

For completeness, we remark that the "maximal type" also exists in general:

Proposition 3.6. *If $\mathfrak{g}$ is semisimple and $\lambda, \mu \in \Lambda^+$, then $m_{\lambda,\mu}^{\lambda+\mu} = 1$. Moreover,*

$$m_{\lambda,\mu}^\nu > 0 \implies \nu \in \mathrm{wt}\, V(\lambda + \mu).$$

More generally, wt $V(\lambda) \otimes V(\mu) = $ wt $V(\lambda + \mu)$.

Inductively, $\operatorname{wt} \otimes_{i=1}^{n} V(\lambda_i) = \operatorname{wt} V(\sum_i \lambda_i)$ if all $\lambda_i \in \Lambda^+$.

Proof. We only show that $\operatorname{wt} V(\lambda) \otimes V(\mu) \subset \operatorname{wt} V(\lambda + \mu)$. Note that $\operatorname{wt} V(\lambda) \otimes V(\mu) = \bigcup_\nu \operatorname{wt} V(\nu)$, where we run over all $\nu \in \Lambda^+$ such that $m_{\lambda,\mu}^\nu > 0$. Now from Kostant's results mentioned above, every such ν is of the form $\lambda + \mu'$, where $\mu' \in \operatorname{wt} V(\mu)$. Hence it suffices to prove that

$$\mu' \in \operatorname{wt} V(\mu), \ \lambda + \mu' \in \Lambda^+ \implies \operatorname{wt} V(\lambda + \mu') \subset \operatorname{wt} V(\lambda + \mu).$$

We now quote a result from [KLV], which says that given $\lambda, \mu \in \Lambda^+, \lambda - \mu \in \mathbb{Z}_{\geq 0} \Pi$ if and only if $\operatorname{conv}(W\mu) \subset \operatorname{conv}(W\lambda)$, where conv denotes the convex hull. Applying this with $\mu \rightsquigarrow \lambda + \mu', \lambda \rightsquigarrow \lambda + \mu$, $\operatorname{conv}(W(\lambda + \mu')) \subset \operatorname{conv}(W(\lambda + \mu))$. Now given $\nu' \in \operatorname{wt} V(\lambda + \mu')$, it is clear that $(\lambda + \mu) - \nu' \in \mathbb{Z}_{\geq 0} \Pi$. Recall [Hal, Theorem 7.41], which says that for all $\lambda \in \Lambda^+$, $\operatorname{wt} V(\lambda) = (\lambda - \mathbb{Z}\Pi) \cap \operatorname{conv}(W\lambda)$. Applying this first with $\lambda \rightsquigarrow \lambda + \mu'$ and then with $\lambda \rightsquigarrow \lambda + \mu$, we get that $\nu' \in \operatorname{conv}(W(\lambda + \mu')) \subset \operatorname{conv}(W(\lambda + \mu))$, so $\nu' \in \operatorname{wt} V(\lambda + \mu)$ as desired. $\square$

3.3. The (K)PRV conjecture and generalized PRV components.

We now discuss a vast generalization of Theorem 3.5, which was conjectured by Parthasarathy et al, extended by Kostant and refined by Verma, and proved by Kumar in [Ku1, Ku3]. The "PRV Conjecture" has been the subject of much study and numerous papers in the literature, and continues to attract interest, as we point out below.

To state the conjecture, recall the following facts from above: given $\lambda, \mu \in \Lambda^+$,

- $m_{\lambda,\mu}^{\lambda + 1 \cdot \mu} = 1$.

- Equation (3.4) says: $\lambda + w_\circ \mu \in \Lambda^+ \implies m_{\lambda,\mu}^{\lambda + w_\circ \mu} = 1$.

There is a common generalization of these assertions to arbitrary $w \in W$, which is mentioned in [Ku5, PY]. Namely, given $\lambda, \mu \in \Lambda^+$ and $w \in W$,

$$\lambda + w\mu \in \Lambda^+ \implies m_{\lambda,\mu}^{\lambda + w\mu} = 1.$$

It is natural to ask what happens when $\lambda + w\mu \notin \Lambda^+$. In light of Theorem 3.5, a natural guess would be to ask if $m_{\lambda,\mu}^{\overline{\lambda + w\mu}} = 1$, or at least, if this multiplicity is positive. This is known as the *PRV Conjecture* in the literature.

Kostant significantly strengthened the PRV conjecture in the following way. Recall that the formal character of each finite-dimensional module $V(\lambda)$ is W-invariant, which implies that for all $w \in W$, $\dim V(\lambda)_{w\lambda} = 1$. Suppose $v_{w\lambda}$ and $v'_{w\mu}$ are nonzero vectors that span the "extremal weight spaces" $V(\lambda)_{w\lambda}$ and $V(\mu)_{w\mu}$ respectively, for all $\lambda, \mu \in \Lambda^+$ and $w \in W$. It is then clear that $v_\lambda \otimes v'_\mu$

generates the copy of the "maximal type" $V(\lambda + \mu)$ inside $V(\lambda) \otimes V(\mu)$. Now consider the minimal type: is it generated by $v_\lambda \otimes v'_{w_\circ \mu}$? The answer is no: in fact, this vector generates the entire module! In other words, $\mathfrak{Ug}(v_\lambda \otimes v'_{w_\circ \mu}) = V(\lambda) \otimes V(\mu)$. Moreover, Theorem 3.5 says that exactly one copy of $V(\overline{\lambda + w_\circ \mu})$ sits in it.

It is now possible to generalize both of these statements. Given any $w \in W$, consider the $\mathfrak{g}$-submodule generated by $v_\lambda \otimes v'_{w\mu}$. Does it contain a (unique) copy of $V(\overline{\lambda + w\mu})$? This is the subject of the *KPRV conjecture*, which was formulated by Kostant and proved by Kumar in [Ku1] in the semisimple case.

Theorem 3.7 ([Ku1, Ku2, Ma1]). *Suppose $\mathfrak{g}$ is semisimple, $\lambda, \mu \in \Lambda^+$, and $w \in W$. Then the module $V(\overline{\lambda + w\mu})$ appears with multiplicity 1 in the submodule $\mathfrak{Ug}(v_\lambda \otimes v'_{w\mu})$ of $V(\lambda) \otimes V(\mu)$.*

Note that a part of Theorem 3.5 is just the special case $w = w_\circ$ of this result. Moreover, the components $V(\overline{\lambda + w\mu})$ are known as *generalized PRV components*.

The KPRV conjecture was also extended to symmetrizable Kac-Moody Lie algebras by Mathieu. Given a symmetrizable generalized Cartan matrix A, one can again define the associated Kac-Moody Lie algebra $\mathfrak{g}(A)$ over a field k. When char $k = 0$, one defines the above notions of dominant integral weights Λ^+ and $\overline{\lambda}$, as well as simple highest weight $\mathfrak{g}(A)$-modules $L(\lambda)$ corresponding to any weight λ. In [Ma1], Mathieu defined an associated Kac-Moody group over an arbitrary field k (after earlier work by Kac, Moody, Peterson, and Tits) using the formalism of ind-schemes. His work led him to prove Theorem 3.7 for $\mathfrak{g}(A)$. (Kumar also proved this case under the assumption that λ is regular, in [Ku2].)

Other proofs of the (K)PRV conjecture have since appeared in the literature (this is from [Ku5]). For example, Polo had proved the PRV conjecture in type A in [Po]. Rajeswari [Ra] gave a proof for classical $\mathfrak{g}$ using Standard Monomial Theory; Littelmann did so using his LS-path model (which generalizes the Littlewood-Richardson rule using tableaux for $\mathfrak{gl}(n)$, to symmetrizable Kac-Moody algebras - see [Li]); and Lusztig's work on the intersection homology of generalized Schubert varieties associated to affine Kac-Moody groups also provides a proof.

Here is another related result. Note that if $w' \leq w$ in the Bruhat order, then $\mathfrak{Ug}(v_\lambda \otimes v'_{w'\mu}) \subset \mathfrak{Ug}(v_\lambda \otimes v'_{w\mu})$. This follows inductively from the case when $w = s_i w' > w'$, which is proved inside the module $V_i((w'\mu)(h_i))$ over the "α_i-copy" of $\mathfrak{sl}_2(\mathbb{C})$, by showing that $v_\lambda \otimes v'_{w'\mu} = e_i^{(w'\mu)(h_i)}(v_\lambda \otimes v'_{w\mu})$. One can now ask if the component $V(\overline{\lambda + w\mu})$ occurs in $\mathfrak{Ug}(v_\lambda \otimes v'_{w'\mu})$ for some $w' < w$, or if $\mathfrak{Ug}(v_\lambda \otimes v'_{w\mu})$ is the "first time" that it occurs in $V(\lambda) \otimes V(\mu)$.

Proposition 3.8 ([Ku1]). *Given regular* $\lambda, \mu \in \Lambda^+$ *(i.e.,* $\lambda(h_i), \mu(h_i) > 0 \,\forall i \in I$*) and* $w' < w$ *in the Bruhat order on* W*, the* $\mathfrak{g}$*-module* $V(\lambda + w\mu)$ *does not occur in* $\mathfrak{Ug}(v_\lambda \otimes v'_{w'\mu})$.

We end this part by mentioning two further directions in which the (original) PRV conjecture has been generalized very recently. Suppose $G \subset \widehat{G}$ are complex connected reductive groups, such that $\widehat{W}$ is the Weyl group of $\widehat{G}$ and various subgroups are "compatible" with the inclusion : $G \hookrightarrow \widehat{G}$ (e.g., $B \subset \widehat{B}$, $T \subset \widehat{T}$, $W \subset \widehat{W}$). Given a dominant integral weight $\widehat{\lambda}$ for $\widehat{G}$ and $\widehat{w} \in \widehat{W}$, does the simple highest-weight (finite-dimensional) module $V_G(\rho(\widehat{w}\widehat{\lambda}))$ with extremal weight $\rho(\widehat{w}\widehat{\lambda})$ occur inside $V_{\widehat{G}}(\widehat{\lambda})$ (upon restricting this to G)? Here, ρ is the restriction of a weight from $\widehat{G}$ to G. For example, the classical PRV conjecture uses $\widehat{G} = G \times G$ containing the diagonal copy of G, and

$$\widehat{W} = W \times W, \qquad \widehat{\lambda} = (\lambda, \mu), \qquad \widehat{w} = (1, w), \qquad \rho(\lambda, \mu) = \lambda + \mu.$$

The above question is addressed in great detail for more general pairs $G \subset \widehat{G}$ in the recent papers [MPR1, MPR2], under the assumption that $\widehat{G}/G$ is "spherical of minimal rank".

Finally, Hayashi has proved a quantum counterpart of the PRV conjecture in [Hay] in the context of fusion rules for $\widehat{\mathfrak{sl}}_3(\mathbb{C})$ and the moduli space of $SU(3)$-flat connections on a pair of pants. These references are intended to reinforce upon the reader that the PRV conjecture is an extremely well-studied result, with connections to several other settings in representation theory and beyond.

3.4. Tensor product multiplicities, revisited.

We now return to the original question in this section, of computing Littlewood-Richardson coefficients. As shown above, several results and formulae have been proposed over the years. Additionally, various other approaches have appeared more recently in the literature. To name but a few: Littelmann's LS-path model, Lusztig's approach using canonical bases, and Kashiwara's use of crystals. See [BZ, Ka, Ku5, Li, Lu] for references and results.

Recall that the basic questions involving tensor product multiplicities are: (a) when are the $m^\nu_{\lambda,\mu}$ positive, and (b) computing the $m^\nu_{\lambda,\mu}$. Results by Kostant, or the PRV conjecture, address the first question, while the PRV theorem and results by Brauer and Steinberg discuss the second one. As the above references and related results show, the work [PRV2] has had quite an influential contribution in this regard.

We conclude this section with a few additional results in this direction (see [Ku5]) that exhibit new components, or in some cases, even obtain a complete decomposition of the tensor product. The first is a refinement of the above PRV conjecture. One can ask if $m^{\overline{\lambda+w\mu}}_{\lambda,\mu} = 1$ for all w, since it is so for $w =$

$1, w_\circ$ from above. This claim turns out to be false - in fact, Verma produced counterexamples for every $\mathfrak{g}$ of rank 2 (i.e., $|\Pi| = 2$), by choosing $\lambda = \mu = \rho = \sum_{i \in I} \varpi_i$.

This led Verma to refine the PRV conjecture as follows. The refined statement was also proved by Kumar.

Theorem 3.9 ([Ku3]). *Given* $\lambda, \mu \in \Lambda^+$ *and* $w \in W$, *define* W_λ *to be the stabilizer subgroup of* λ *in* W, *and the map* $\eta : W_\lambda \backslash W / W_\mu \to \Lambda^+$ *via:* $\eta(W_\lambda w W_\mu) = \overline{\lambda + w\mu}$. *Then* $m_{\lambda,\mu}^{\overline{\lambda+w\mu}} \geq \#\eta^{-1}(\eta(W_\lambda w W_\mu))$.

Of course, if λ, μ are both regular (i.e., (λ, α) and (μ, α) are both nonzero for all roots α), then $W_\lambda = W_\mu = \{1\}$.

Second, a related result from [Ku3] is able to determine *all* the multiplicities when wt $V(\mu) = W\mu$ is a single orbit (i.e., μ is minuscule). In this case,

$$V(\lambda) \otimes V(\mu) \cong \bigoplus_{\overline{w} \in W/W_\mu : \lambda + w\mu \in \Lambda^+} V(\lambda + w\mu),$$

where each factor occurs with multiplicity 1. There are exactly $\#W_\lambda \backslash W / W_\mu$ components.

Kumar also shows the following result in [Ku4]: suppose $\beta \in R^+$ is such that $\lambda + \mu - \beta \in \Lambda^+$, and such that $\beta - \alpha_i \notin R^+ \cup \{0\}$ whenever $\lambda(h_i)$ or $\mu(h_i) = 0$. Then $m_{\lambda,\mu}^{\lambda+\mu-\beta} > 0$. A similar result can be found in the recent work [MPR1], where the authors demonstrate new components of the form $w_1\lambda + w_2\mu - k\alpha_i$ for some $w_1, w_2 \in W$ and $i \in I$.

Finally, Dimitrov and Roth have worked with restrictions of line bundles from the square $G/B \times G/B$ of the flag variety to the diagonally embedded copy. (Here, G is a connected reductive algebraic group with $\mathrm{Lie}(G) = \mathfrak{g}$, and, $B \subset G$ is a Borel subgroup with $\mathrm{Lie}(B) = \mathfrak{h} \oplus \mathfrak{n}^+$; Kumar's proofs in [Ku1] involved a study of similar objects.) They study special components $V(\nu)$ of the tensor product modules $V(\lambda) \otimes V(\mu)$, that arise out of cohomological reasons. The authors mention in [DR] that these *cohomological components* automatically turn out to be generalized PRV components satisfying: $m_{k\lambda,k\mu}^{k\nu} = 1$ for all $k \in \mathbb{N}$. They go on to prove the converse implication when G is a classical group, as well as in other cases.

4. Irreducible Banach Space Representations

We now continue the discussion prior to the preceding section, about constructing irreducible $\overline{\mathfrak{g}}$-admissible Banach (actually, Hilbert) space representations of G. In [PRV2], having proved that finite-dimensional $\mathfrak{g} \times \mathfrak{g}$-modules have minimal types, the authors proceed to construct other such representations (possibly infinite-dimensional), in a completely different manner. Their construction depends heavily on the work of Harish-Chandra [Har1]–[Har4]. These representations $\widehat{\pi}_{\lambda,\nu}$ are defined on subquotients of a Hilbert space.

4.1. The work of Harish-Chandra.

In order to outline the construction of these $\mathfrak{g}$-modules by Harish-Chandra and by Parthasarathy et al, additional notation is needed. Let $K \subset G$ be the maximal compact subgroup of a complex connected semisimple Lie group, and let $\mathfrak{h}_0 \subset \text{Lie}(K)$ be a Cartan subalgebra. Now define $M := \exp(\sqrt{-1} \cdot \mathfrak{h}_0) \subset K$ to be the corresponding Cartan subgroup. For each $\nu \in \Lambda$, define σ_ν to be the unique character of M that sends $\exp(\sqrt{-1} \cdot h)$ to $\exp(\sqrt{-1} \cdot \nu(h))$ for all $h \in \mathfrak{h}_0$. Then $\nu \mapsto \sigma_\nu$ is an isomorphism of $(\Lambda, +)$ onto the character group $\widehat{M}$ of M.

Now let $\mathfrak{H} := L^2(K, \mathbb{C}, \mu)$, where μ denotes the (normalized) Haar measure on the compact group K. This is a representation of M under the right-regular action: $(m \cdot f)(k) := f(km^{-1})$. Given $\nu \in \Lambda$, define the ν-weight subspace of $\mathfrak{H}$ as follows:

$$\mathfrak{H}(\nu) := \{f \in \mathfrak{H} : m \cdot f = \sigma_{-\nu}(m)f \ \forall m \in M\}.$$

Then $\mathfrak{H}$ decomposes as the direct sum of the $\mathfrak{H}(\nu)$ over all $\nu \in \Lambda$. Moreover, given $\xi \in \mathfrak{h}^*$, Harish-Chandra had previously defined and studied a G-module structure π_ξ on $\mathfrak{H}$ in [Har2]–[Har4]. It turns out that every $\mathfrak{H}(\nu)$ is a submodule of $\mathfrak{H}$ under this structure; define $\pi_{\xi,\nu}$ to be this representation. Here are some of the properties of these modules that are used in [PRV2].

Theorem 4.1 (Harish-Chandra). *Fix $\xi \in \mathfrak{h}^*$ and $\nu \in \Lambda$.*

1. *For all $\mu \in \Lambda^+$, $[\pi_{\xi,\nu} : V_{\overline{\mathfrak{g}}}(\mu)] = \dim V_{\overline{\mathfrak{g}}}(\mu)_\nu$. In particular, $[\pi_{\xi,\nu} : V_{\overline{\mathfrak{g}}}(\overline{\nu})] = 1$.*

2. *The representation $\pi_{\xi,\nu}$ has an infinitesimal character (of $Z(\mathfrak{U}(\mathfrak{g} \times \mathfrak{g}))$).*

3. *$\pi_{\xi,\nu}$ possesses a distributional character $\Theta_{\xi,\nu}$, which is a locally summable function that is analytic on the dense open subset of regular points of G. Moreover, $\Theta_{\xi,\nu} = \Theta_{\xi',\nu'}$ if and only if there exists $w \in W$ such that $\xi' = w\xi, \nu' = w\nu$.*

4. *If ξ is restricted to lie in the real subspace $\mathfrak{R}$ of all weights that take purely imaginary values on $\mathfrak{h}_0$, then $\pi_{\xi,\nu}$ is always unitary, and almost always irreducible, say whenever $\xi \in \mathfrak{R}_\nu \subset \mathfrak{R}$ (for each ν). In particular, if $\xi \in \mathfrak{R}_\nu$ and $w \in W$, then $\pi_{\xi,\nu} \cong \pi_{w\xi,w\nu}$.*

Although it is not specifically mentioned in [PRV2], it is actually possible to compute the central character of $\pi_{\xi,\nu}$ - and this has a very familiar expression. See Section 4.7.

4.2. Constructing the representations $\widehat{\pi}_{\lambda,\nu}$.

For his subquotient theorem, Harish-Chandra identified two closed subspaces $\mathfrak{H}''_\xi(\nu) \subset \mathfrak{H}'_\xi(\nu) \subset \pi_{\xi,\nu}$, such that the quotient of the larger of them by the smaller one is an irreducible G-module. Obtaining a greater understanding of these subquotients of $\pi_{\xi,\nu}$ was one of the main motivations behind [PRV2]; when G is a complex semisimple group, the authors are indeed able to describe these subspaces more easily than Harish-Chandra in the real case. We start with this description. The remainder of this entire section is based on [PRV2, Section 2.4].

Recall from Theorem 4.1 that $[\pi_{\xi,\nu} : V_{\overline{\mathfrak{g}}}(\overline{\nu})] = 1$. Thus, define $\mathfrak{H}'_\xi(\nu)$ to be the smallest closed G-submodule of $\pi_{\xi,\nu}$ containing the unique copy of $V_{\overline{\mathfrak{g}}}(\overline{\nu})$. Inside this, define $\mathfrak{H}''_\xi(\nu)$ to be the sum of all closed G-submodules $M \subset \mathfrak{H}'_\xi(\nu)$ such that $M \cap V_{\overline{\mathfrak{g}}}(\overline{\nu}) = 0$. Then $\mathfrak{H}''_\xi(\nu)$ is a maximal submodule of $\mathfrak{H}'_\xi(\nu)$, and this leads to the irreducible G-representations $\widehat{\pi}_{\lambda,\nu} := \mathfrak{H}'_\xi(\nu)/\mathfrak{H}''_\xi(\nu)$, where $\lambda := \frac{1}{2}(\xi+\nu) - \rho$ runs over all of $\mathfrak{h}^*$ as well. Clearly, $[\widehat{\pi}_{\lambda,\nu} : V_{\overline{\mathfrak{g}}}(\mu)] \leq \dim V_{\overline{\mathfrak{g}}}(\mu)_\nu \ \forall \mu \in \Lambda^+$.

The space $\widehat{\pi}_{\lambda,\nu}$ was shown in [PRV2] to have the following properties:

- $\widehat{\pi}_{\lambda,\nu}$ is an irreducible subquotient of $\pi_{\xi,\nu} \subset \mathfrak{H} = L^2(K,\mathbb{C},\mu)$, hence it too is defined on a Hilbert space. Moreover, $\pi_{\xi,\nu}$ is irreducible if and only if $\widehat{\pi}_{\lambda,\nu} \cong \pi_{\xi,\nu}$, if and only if $[\widehat{\pi}_{\lambda,\nu} : V_{\overline{\mathfrak{g}}}(\mu)] = \dim V_{\overline{\mathfrak{g}}}(\mu)_\nu$ for all $\mu \in \Lambda^+$.

- $\widehat{\pi}_{\lambda,\nu}$ is an object of $\mathcal{C}(\mathfrak{g} \times \mathfrak{g}, \overline{\mathfrak{g}})$, with minimal type component $\overline{\nu} \in \Lambda^+ \cap W\nu$. Moreover, $[\widehat{\pi}_{\lambda,\nu} : V_{\overline{\mathfrak{g}}}(\overline{\nu})] = 1$.

- $\widehat{\pi}_{\lambda,\nu}$ has the same infinitesimal character as $\pi_{\xi,\nu}$, where $\lambda = \frac{1}{2}(\xi+\nu) - \rho$.

Note that if $\nu' \notin W\nu$, then $\widehat{\pi}_{\lambda',\nu'}$ and $\widehat{\pi}_{\lambda,\nu}$ cannot be isomorphic by Equation (2.8), because their minimal types are $\overline{\nu'} \neq \overline{\nu}$ respectively.

From above, the modules $\widehat{\pi}_{\lambda,\nu}$ admit infinitesimal characters. It is clear that the highest weight modules $V(\lambda,\mu)$ also admit such characters. Moreover, both of these are families of simple objects in $\mathcal{C}(\mathfrak{g} \times \mathfrak{g}, \overline{\mathfrak{g}})$. Therefore it is natural to ask if $\widehat{\pi}_{\lambda,\nu}$ is finite-dimensional for some values of the parameters - and if all finite-dimensional simple modules $V(\lambda,\mu)$ are thus covered.

To answer these questions (affirmatively!), Parthasarathy et al studied the "key homomorphisms" $\eta_{V,\mathcal{D}} : \Omega \to \mathbb{C}$ in greater detail, by relating them to certain homomorphisms $\mathbf{h}^{\Pi} : \Omega \to P(\mathfrak{h}^* \times \mathfrak{h}^*)$. These homomorphisms are the subject of the next subsection.

4.3. Constructing the polynomial-valued maps $\mathbf{h}^{\Pi'}$.

Recall the "key homomorphism" $\eta_{V,\mathcal{D}} : \Omega \to \mathbb{C}$, that is defined whenever a simple $\overline{\mathfrak{g}}$-module $\mathcal{D}$ arises with multiplicity one in a simple object V of $\mathcal{C}(\mathfrak{g} \times \mathfrak{g}, \overline{\mathfrak{g}})$. It turns out that there is an explicit construction of the map $\eta_{\widehat{\pi}_{\lambda,\nu},\overline{\nu}}$ via a different homomorphism $\mathbf{h}^{\Pi'}(-;\lambda,\nu)$, which we now present. We explicitly compute both of these maps below in the example of $\mathfrak{g} = \mathfrak{sl}_2(\mathbb{C})$, to show that they are equal. This material discusses [PRV2, Sections 2.3, 2.4].

To construct the map $\mathbf{h}^{\Pi'}$, some more notation is needed. Given $X \in \mathfrak{g}$, define

$$X_{(1)} := X \otimes 1, \qquad X_{(2)} := 1 \otimes X, \qquad \overline{X} := X_{(1)} + X_{(2)},$$

and similarly, $\mathfrak{g}_{(1)}, \mathfrak{g}_{(2)} \subset \widehat{\mathfrak{g}} = \mathfrak{g} \oplus \mathfrak{g}$, as well as $\mathfrak{h}_{(1)}$ and so on. Then $X \mapsto \overline{X}$ extends to an isomorphism of associative algebras : $\mathfrak{Ug} \to \mathfrak{U\overline{g}}$, and similar statements hold for $\mathfrak{g}_{(2)}, \mathfrak{h}_{(1)}$, etc.

Now define $\widehat{\mathfrak{q}} := \mathfrak{n}_{(1)}^+ \oplus \mathfrak{n}_{(2)}^-$. Note that this is the "positive part" of the triangular decomposition of $\mathfrak{g} \times \mathfrak{g}$, if we define $\widehat{\Pi} := \Pi_{(1)} \coprod -\Pi_{(2)}$. This choice of simple roots for $\widehat{\mathfrak{g}}$ comes from the consideration of the conjugation on $\mathfrak{g}$ with respect to a compact form; see a previous footnote. Now $\widehat{\mathfrak{g}} = \overline{\mathfrak{g}} \oplus \mathfrak{h}_{(1)} \oplus \widehat{\mathfrak{q}}$, so by the PBW Theorem,

$$\mathfrak{U\widehat{g}} \cong (\mathfrak{U\overline{g}} \otimes \mathfrak{Uh}_{(1)}) \oplus (\mathfrak{U\widehat{g}})\widehat{\mathfrak{q}}$$

as $\mathbb{C}$-vector spaces. Note that every $H \in \operatorname{Sym}\mathfrak{h}$ is a polynomial on $\mathfrak{h}^*$ as follows: write $H = p(\{h_i : i \in I\})$ for some polynomial p. Then $H(\lambda) = p(\{\lambda(h_i) : i \in I\})$. This also applies to $H \in \operatorname{Sym}\mathfrak{h}_{(1)}$ or $\operatorname{Sym}\overline{\mathfrak{h}}$, for instance, via the obvious isomorphisms mentioned above. Similarly, define $\mathbf{h}^{\mathbf{n}} := \prod_{i \in I} h_i^{n_i}$ for $\mathbf{n} = (n_i)_{i \in I} \in \mathbb{Z}_{\geq 0}^I$ (which we write as: $\mathbf{n} \geq \mathbf{0}$), and also $\overline{\mathbf{h}}^{\mathbf{n}}, \mathbf{h}_{(1)}^{\mathbf{n}}$, and $\lambda(\mathbf{h})^{\mathbf{n}} = \lambda(\mathbf{h}^{\mathbf{n}}) := \mathbf{h}^{\mathbf{n}}(\lambda)$ from above.

We can now define the homomorphisms in question. Suppose $\omega \in \Omega$, the centralizer of $\overline{\mathfrak{g}}$ in $\mathfrak{U\widehat{g}}$. Then there exists a unique $\xi_{\mathbf{n}} \in \mathfrak{Ug}$ for all $\mathbf{n} \geq \mathbf{0}$, such that

$$\omega \equiv \sum_{\mathbf{n} \geq \mathbf{0}} \overline{\xi_{\mathbf{n}}} \otimes \mathbf{h}_{(1)}^{\mathbf{n}} \quad \mod (\mathfrak{U\widehat{g}})\widehat{\mathfrak{q}}.$$

Since $[\overline{h}, \omega] = 0$ for all $h \in \mathfrak{h}$, one checks that $\xi_{\mathbf{n}} \in (\mathfrak{Ug})_0$ for all $\mathbf{n}$. Finally, given any subset $\Pi' \subset R$ of simple roots for some Borel subalgebra (equivalently, $\Pi' = w\Pi$ for some $w \in W$), the maps $\mathbf{h}^{\Pi'}$ are defined as follows:

$$\mathbf{h}^{\Pi'}(\omega) := \sum_{\mathbf{n} \geq \mathbf{0}} \beta^{\Pi'}(\xi_{\mathbf{n}}) \otimes \mathbf{h}^{\mathbf{n}} \in \operatorname{Sym}(\mathfrak{h} \times \mathfrak{h}),$$

$$\mathbf{h}^{\Pi'}(\omega; \lambda, \nu) := \sum_{\mathbf{n} \geq \mathbf{0}} \nu(\beta^{\Pi'}(\xi_{\mathbf{n}}))\lambda(\mathbf{h}^{\mathbf{n}}) \; \forall \lambda, \nu \in \mathfrak{h}^*.$$

It turns out that these polynomials are very familiar expressions, when ω is restricted to lie in the center $Z(\mathfrak{U}(\mathfrak{g} \times \mathfrak{g}))$. We see this in Section 4.7 below.

4.4. Relationship between $\widehat{\pi}_{\lambda,\nu}$ and $\mathbf{h}^{\Pi'}(-;\lambda,\nu)$.

Recall that there are two classes of irreducible admissible representations that are constructed in [PRV2]: the finite-dimensional modules $V(\lambda,\mu)$ for $\lambda,\mu \in \Lambda^+$, and the Hilbert space representations $\widehat{\pi}_{\lambda,\nu}$ for $\lambda \in \mathfrak{h}^*$ and $\nu \in \Lambda$. In the former case, we define $\nu := \lambda + w_\circ\mu \in \Lambda$ (as in Theorem 3.5); then in both families, the representations all contain the minimal type $V_{\overline{\mathfrak{g}}}(\overline{\nu})$ with multiplicity 1.

Now how does one show that the first of the above families is actually contained inside the second? Similarly, how does one check if two given representations $\widehat{\pi}_{\lambda,\nu}$ and $\widehat{\pi}_{\lambda',\nu'}$ are equivalent or not? The answer in both cases is to use the homomorphisms $\mathbf{h}^{\Pi'}$, together with Equation (2.8). More precisely, one relates the maps $\mathbf{h}^{\Pi'}$ to the homomorphisms $\eta_{\widehat{\pi}_{\lambda,\nu},\overline{\nu}}$. (Note that this does not completely answer the second question.)

Here are some results from the heart of [PRV2], in which the authors begin to address these questions. The proofs use Theorem 4.1.

Theorem 4.2 ([PRV2]). *Suppose $\lambda \in \mathfrak{h}^*$ and $w(\nu) \in \Lambda^+$ for some $\nu \in \Lambda, w \in W$.*

1. *Then $\eta_{\widehat{\pi}_{\lambda,\nu},\overline{\nu}}(-) \equiv \mathbf{h}^{w^{-1}\Pi}(-;\lambda,\nu)$ are homomorphisms : $\Omega \to \mathbb{C}$.*

2. *The maps $\mathbf{h}^{w\Pi}$ are homomorphisms : $\Omega \to P(\mathfrak{h}^* \times \mathfrak{h}^*)$ for all $w \in W$. They are W-equivariant in the following sense: for all $\omega \in \Omega$, $w, w' \in W$, and $\lambda, \nu \in \mathfrak{h}^*$,*

$$\mathbf{h}^{w'w\Pi}(\omega; w' * \lambda, w'\nu) = \mathbf{h}^{w\Pi}(\omega; \lambda, \nu).$$

A consequence of this result is a "first step" towards the classification of the representations $\widehat{\pi}_{\lambda,\nu}$. (This is discussed at greater length in Section 7.)

Corollary 4.3. *Suppose $(\lambda,\nu), (\lambda',\nu') \in \mathfrak{h}^* \times \Lambda$. Then $\widehat{\pi}_{\lambda,\nu} \cong \widehat{\pi}_{w*\lambda,w\nu}$ for all $w \in W$, while $\widehat{\pi}_{\lambda,\nu}$ and $\widehat{\pi}_{\lambda',\nu'}$ are not equivalent if $\nu' \notin W\nu$.*

Remark 4.4. Another consequence is the following. Note that since every Verma module has a unique simple quotient, hence there exists a unique maximal (left) ideal $\mathfrak{M}_\lambda \subset \mathfrak{U}\mathfrak{g}$ containing $\mathfrak{n}^+$ and ker λ. If $\lambda \in \Lambda^+$, then by [Har1],

$$\mathfrak{M}_\lambda = (\mathfrak{U}\mathfrak{g})\mathfrak{n}^+ + (\mathfrak{U}\mathfrak{g})\ker\lambda + \sum_{i\in I} \mathfrak{U}\mathfrak{g} \cdot f_i^{\lambda(h_i)+1}.$$

Thus whenever $w(\nu) \in \Lambda^+$ for $\nu \in \Lambda$ and $w \in W$, there exists a unique maximal ideal in $\mathfrak{U}\overline{\mathfrak{g}}$ containing ker $\nu \subset \overline{\mathfrak{h}}$ and $\{\overline{e_\alpha} : \alpha \in w^{-1}(R^+)\}$, where e_α spans $\mathfrak{g}_\alpha$. Call this ideal $\overline{\mathfrak{M}}_\nu$. Now since $\widehat{\pi}_{\lambda,\nu}$ is a simple $\mathfrak{g} \times \mathfrak{g}$-module, it is generated by the $\overline{\mathfrak{g}}$-(maximal) weight vector $v_{\overline{\nu}} \in V_{\overline{\mathfrak{g}}}(\overline{\nu})_{\overline{\nu}} \subset \widehat{\pi}_{\lambda,\nu}$. Moreover, Ω acts on $V_{\overline{\mathfrak{g}}}(\overline{\nu})$

by $\mathbf{h}^{w^{-1}\Pi}(-;\lambda,\nu)$. By Equation (2.8), this data uniquely determines $\widehat{\pi}_{\lambda,\nu}$ up to isomorphism. This implies the following result from [PRV1]:

There exists a unique maximal ideal $\mathfrak{M}_{\lambda,\nu} \subset \mathfrak{U}(\mathfrak{g} \times \mathfrak{g})$ *containing* $\ker \mathbf{h}^{w^{-1}\Pi}(-;\lambda,\nu) \subset \Omega$ *and* $\overline{\mathfrak{M}}_{\nu}$*. Moreover,* $\widehat{\pi}_{\lambda,\nu} \cong \mathfrak{U}(\mathfrak{g} \times \mathfrak{g})/\mathfrak{M}_{\lambda,\nu}$*.*

Note that $\mathbf{h}^{\Pi'} : \Omega \to P(\mathfrak{h}^* \times \mathfrak{h}^*)$ is a homomorphism for each $\Pi' = w\Pi$. Thus, $\mathbf{h}^{\Pi'}(-;\lambda,\nu)$ is a homomorphism : $\Omega \to \mathbb{C}$ for all $\lambda, \nu \in \mathfrak{h}^*$. The strategy in [PRV2] for showing this is to restrict to the Zariski dense subset of (λ, ν) arising from finite-dimensional modules. The authors prove that if μ and Π' are chosen "suitably", then $\mathbf{h}^{\Pi'}(-;\lambda,\nu) \equiv \eta_{V(\lambda,\mu),\overline{\nu}}(-)$ is a homomorphism : $\Omega \to \mathbb{C}$. Hence so is $\mathbf{h}^{\Pi'}$ at all values of (λ,ν). This analysis leads to the next topic.

4.5. Relationship between $V(\lambda,\mu)$ and $\mathbf{h}^{\Pi'}(-;\lambda,\nu)$.

Consider the other family of simple $\mathcal{C}(\mathfrak{g} \times \mathfrak{g}, \overline{\mathfrak{g}})$-modules studied in [PRV2]: the finite-dimensional $V(\lambda,\mu)$. From above, the minimal type of such a module is $\overline{\lambda + w_\circ\mu}$. Now consider the "converse" question: given $\lambda \in \mathfrak{h}^*$ and $\nu \in \Lambda$, is it possible to produce $\mu \in \Lambda^+$ such that $V(\lambda,\mu)$ has minimal type $\overline{\nu}$? Supporting evidence for such a claim is given by the following result, in light of Equation (2.8).

Theorem 4.5 ([PRV2]). *Suppose* $\lambda \in \Lambda^+$ *and* $\nu \in \lambda - \Lambda^+$*. Choose* $w \in W$ *such that* $w(\nu) \in \Lambda^+$ *and define* $\mu := -w_\circ(\lambda - \nu) \in \Lambda^+$*. Then* $V(\lambda,\mu)$ *has minimal type* $V_{\overline{\mathfrak{g}}}(\overline{\nu})$*; moreover,*

$$\eta_{V(\lambda,\mu),\overline{\nu}}(-) \equiv \mathbf{h}^{w^{-1}\Pi}(-;\lambda,\nu).$$

Restating this allows us to answer the "converse" question, which can also be found in [Du2].

Corollary 4.6. *For all* $\lambda, \mu \in \Lambda^+$*,* $V(\lambda,\mu) \cong \widehat{\pi}_{\lambda,\lambda+w_\circ\mu}$*.*

Proof. Note that $\nu = \lambda + w_\circ\mu \in \lambda - \Lambda^+ \Leftrightarrow \mu = -w_\circ(\lambda - \nu) \in \Lambda^+$. Thus, set $\nu := \lambda + w_\circ\mu$ and choose $w \in W$ such that $w(\nu) \in \Lambda^+$. Then using Theorems 4.2 and 4.5,

$$\eta_{\widehat{\pi}_{\lambda,\lambda+w_\circ\mu},\overline{\nu}}(-) \equiv \mathbf{h}^{w^{-1}\Pi}(-;\lambda,\lambda+w_\circ\mu) \equiv \eta_{V(\lambda,\mu),\overline{\nu}}(-)$$

on Ω. Moreover, $\overline{\nu}$ is the (multiplicity one) minimal type component of both irreducible modules, by the PRV Theorem. The result follows by applying Equation (2.8). $\square$

Remark 4.7. A word of caution: note that $V(\lambda) \otimes V(\mu) \cong V(\mu) \otimes V(\lambda)$ as $\overline{\mathfrak{g}}$-modules for $\lambda, \mu \in \mathfrak{h}^*$, since $\mathfrak{U}\mathfrak{g}$ is cocommutative. Thus, their minimal types are also equal (when $\lambda, \mu \in \Lambda^+$): $\overline{\lambda + w_o\mu} = \overline{\mu + w_o\lambda}$. This implies that the action of $Z(\mathfrak{U}\overline{\mathfrak{g}})$ is equal on both modules. However, $V(\lambda, \mu)$ and $V(\mu, \lambda)$ are non-isomorphic simple $\mathfrak{g} \times \mathfrak{g}$-modules if $\lambda \neq \mu$. Similarly, the infinitesimal characters are not equal on all of $Z(\mathfrak{U}(\mathfrak{g} \times \mathfrak{g}))$ - and hence the key homomorphisms $\eta_{V(\lambda,\mu),\overline{\nu}}, \eta_{V(\mu,\lambda),\overline{\nu}}$ do not agree on all of Ω - unless $\lambda = \mu$.

4.6. Example: the case of $\mathfrak{sl}_2(\mathbb{C})$.

We now verify a part of Theorem 4.5 in the special case of $\mathfrak{g} = \mathfrak{sl}_2(\mathbb{C})$, in which case it is not hard to compute both the homomorphisms in question - at least, on a particular finitely generated subalgebra $\Omega' \subset \Omega$.

For convenience, denote the two factors in $\widehat{\mathfrak{g}}$ as $\mathfrak{g}_k$ for $k = 1, 2$ (as opposed to $\mathfrak{g}_{(1)}$ and $\mathfrak{g}_{(2)}$), with bases $\{e_k, f_k, h_k\}$. Now for all $0 < n_2 \leq n_1 \in \mathbb{N}$, the "tensor product" module $V_1(n_1) \otimes V_2(n_2)$ is a simple object in $\mathcal{C}(\mathfrak{g} \times \mathfrak{g}, \overline{\mathfrak{g}})$, of highest weight (n_1, n_2). Restricted to $\overline{\mathfrak{g}}$, it decomposes according to the Clebsch-Gordan coefficients:

$$V(n_1, n_2) \cong_{\overline{\mathfrak{g}}} V(n_1) \otimes V(n_2) \cong V_{\overline{\mathfrak{g}}}(n_1+n_2) \oplus V_{\overline{\mathfrak{g}}}(n_1+n_2-2) \oplus \cdots \oplus V_{\overline{\mathfrak{g}}}(n_1-n_2),$$

where $V_{\overline{\mathfrak{g}}}(n)$ is a finite-dimensional irreducible $\overline{\mathfrak{g}}$-module of highest weight n (equivalently, of dimension $n + 1$). Now denote the highest weight generators of $V_i(n_i)$ by v_{n_1} and v'_{n_2} respectively, and define the weight basis $v_{n_1-2i} := (f^i/i!)v_{n_1}$ of $V_1(n_1)$, with $0 \leq i < \dim V_1(n_1)$. Similarly define $v'_{n_2-2j} \in V_2(n_2)$. One checks using Equation (1.1) that

$$v''_{n_1-n_2} := \sum_{j=0}^{n_2} (-1)^j j! \cdot n_1(n_1 - 1) \cdots (n_1 - j + 1) \cdot v_{n_1-2j} \otimes v'_{2j-n_2}$$

is a weight vector in $V(n_1, n_2)$ which is killed by $e_1 + e_2 := e_1 \otimes 1 + 1 \otimes e_2$. Hence it generates the minimal type (i.e., the PRV component) $V_{\overline{\mathfrak{g}}}(n_1 - n_2)$, and we have:

$$\lambda = n_1, \qquad \mu = n_2, \qquad \nu = \lambda + w_o\mu = n_1 - n_2 \geq 0, \qquad \overline{\nu} = \nu \in \Lambda^+.$$

Note that the subalgebra Ω that commutes with $\overline{\mathfrak{g}}$ in $\mathfrak{U}(\widehat{\mathfrak{g}})$ contains the center of $\mathfrak{U}(\widehat{\mathfrak{g}})$. This is freely generated by the two Casimir operators

$$\Delta_i := 4f_ie_i + h_i^2 + 2h_i = 4e_if_i + h_i^2 - 2h_i.$$

Clearly, $\Delta_1 \otimes 1$ acts on $V_1(n_1) \otimes V_2(n_2)$ by the scalar $n_1^2 + 2n_1$; similarly, $1 \otimes \Delta_2$ acts by $n_2^2 + 2n_2$. Moreover, $\overline{\Delta} = 4(f_1 + f_2)(e_1 + e_2) + (h_1 + h_2)^2 + 2(h_1 + h_2)$ lies in Ω as well; by $\mathfrak{sl}_2$-theory, it acts on $v''_{n_1-n_2}$ via: $\overline{\Delta} \cdot v''_{n_1-n_2} = ((n_1 - n_2)^2 + 2(n_1 - n_2))v''_{n_1-n_2}$. Since $\overline{\Delta}$ commutes with $\overline{\mathfrak{g}}$, it acts on $V_{\overline{\mathfrak{g}}}(n_1 - n_2)$ by

110

this same scalar. This allows us to determine $\eta_{V(n_1,n_2),n_1-n_2}$ - at least, on the subalgebra $\Omega' := \mathbb{C}[\Delta_1, \Delta_2, \overline{\Delta}] \subset \Omega$:

$$\eta_{V(n_1,n_2),n_1-n_2}(\Delta_i) := n_i^2 + 2n_i = \chi(n_i)(\Delta) \ (i = 1, 2),$$
$$\eta_{V(n_1,n_2),n_1-n_2}(\overline{\Delta}) := \chi(n_1 - n_2)(\Delta). \tag{4.8}$$

Now consider $\mathbf{h}^\Pi(-;\lambda,\nu)$. Note from above that $\overline{\nu} = \nu = \lambda + w_\circ\mu = n_1 - n_2$, so $w = 1$. Since $\widehat{\mathfrak{q}}$ is spanned by e_1 and f_2, projecting the various Δ_i onto $\mathfrak{U}\overline{\mathfrak{g}} \otimes \mathfrak{U}\mathfrak{h}_{(1)}$ modulo $\widehat{\mathfrak{q}}$ yields

$$\Delta_1 = 4f_1e_1 + h_1^2 + 2h_1 \equiv h_1^2 + 2h_1 \quad \mathrm{mod} \ (\mathfrak{U}\widehat{\mathfrak{g}})\widehat{\mathfrak{q}}.$$

Hence $\mathbf{h}^\Pi(\Delta_1;\lambda,\nu) = \lambda(h_1)^2 + 2\lambda(h_1) = n_1^2 + 2n_1$, as in Equation (4.8). Similarly,

$$\begin{aligned}\Delta_2 &= 4e_2f_2 + h_2^2 - 2h_2 \equiv h_2^2 - 2h_2 \equiv (h_1 + h_2)^2 - 2(h_1 + h_2) - h_1^2 - 2h_1h_2 + 2h_1 \\ &\equiv \overline{h}^2 - 2\overline{h} - 2\overline{h} \otimes h_1 + h_1^2 + 2h_1 \quad \mathrm{mod} \ (\mathfrak{U}\widehat{\mathfrak{g}})\widehat{\mathfrak{q}}.\end{aligned}$$

Computing $\mathbf{h}^\Pi(\Delta_2;\lambda,\nu)$ amounts to evaluating this polynomial at $(\nu(\overline{h}),\lambda(h_1)) = (n_1 - n_2, n_1)$. This yields $n_2^2 + 2n_2$, as desired. Finally, $\overline{\Delta} = 2\overline{f}\overline{e} + \overline{h}^2 + 2\overline{h}$. To evaluate $\mathbf{h}^\Pi(\overline{\Delta};\lambda,\nu)$, we must first apply the Harish-Chandra projection β^Π to this expression, which kills the first term. Now evaluating at $(\overline{h}, h_1) = (n_1 - n_2, n_1)$, we obtain $\chi(n_1 - n_2)(\Delta)$, as in Equation (4.8). Thus, $\mathbf{h}^\Pi(\Delta';n_1,n_1-n_2) \equiv \eta_{V(n_1,n_2),n_1-n_2}(\Delta')$ for $\Delta' = \Delta_1, \Delta_2, \overline{\Delta}$; assuming that both of these are homomorphisms implies equality on all of Ω'. $\qquad\square$

4.7. Infinitesimal characters.

Although it does not seem to be explicitly mentioned in [PRV2], the above facts allow us to compute the infinitesimal characters of the representations $\pi_{\xi,\nu}$ - or equivalently, of $\widehat{\pi}_{\lambda,\nu}$. In fact, we prove a stronger result.

Theorem 4.9. *For all $\lambda, \nu \in \mathfrak{h}^*$, $w \in W$, and $z \in Z(\mathfrak{U}(\mathfrak{g} \times \mathfrak{g}))$,*

$$\mathbf{h}^{w\Pi}(z;\lambda,\nu) = \chi(\lambda, \nu - \lambda - 2\rho)(z),$$

where $\chi(\lambda, \lambda')$ is the central character for the $\mathfrak{g} \times \mathfrak{g}$-Verma module $M(\lambda, \lambda')$. Now given $\xi \in \mathfrak{h}^*$ and $\nu \in \Lambda$, define $\lambda := \frac{1}{2}(\xi + \nu) - \rho$. Then $\chi_{\pi_{\xi,\nu}} = \chi_{\widehat{\pi}_{\lambda,\nu}} = \chi(\lambda, \nu - \lambda - 2\rho)$.[5] Moreover, $\chi_{\pi_{w\xi,w\nu}} = \chi_{\widehat{\pi}_{w*\lambda,w\nu}} = \chi(w*\lambda, w*(\nu - \lambda - 2\rho)) = \chi(\lambda, \nu - \lambda - 2\rho)$.

---

[5]This assertion about central characters of Harish-Chandra modules can be found in [Du2], for instance.

For example, consider the situation where $\widehat{\pi}_{\lambda,\nu}$ is finite-dimensional. Thus, $\nu \in \lambda - \Lambda^+$, and $\widehat{\pi}_{\lambda,\nu} = V(\lambda, \mu)$, where $\mu := -w_\circ(\lambda - \nu) \in \Lambda^+$. Thus, $\nu = \lambda + w_\circ\mu$. In this case, the result says that the second component of the central character is:

$$\chi(\nu - \lambda - 2\rho) = \chi(\lambda + w_\circ\mu - \lambda - 2\rho) = \chi(w_\circ\mu + w_\circ\rho - \rho) = \chi(w_\circ * \mu),$$

since $w_\circ\rho = -\rho$. Hence $\chi_{V(\lambda,\mu)} = \chi(\lambda, w_\circ * \mu) = \chi(\lambda, \mu)$ (by Theorem 1.2), as expected.

Proof. We use a "Zariski density" argument as in [PRV2], that Varadarajan attributes to Harish-Chandra in [Va]. By Harish-Chandra's Theorem 1.2, every central character of $Z(\mathfrak{U}(\mathfrak{g} \times \mathfrak{g}))$ is of the form $\chi(\mu_1, \mu_2)$ for some $\mu_i \in \mathfrak{h}^*$. Moreover, $Z(\mathfrak{U}(\mathfrak{g} \times \mathfrak{g})) \subset \Omega$, so we can evaluate $\mathbf{h}^{w^{-1}\Pi}(-; \lambda, \nu)$ on it. For all $z \in Z(\mathfrak{U}\mathfrak{g})$, one uses the definitions to check that $\mathbf{h}^{w^{-1}\Pi}(z_{(1)}; \lambda, \nu) = (\nu \otimes \lambda)(1 \otimes \beta^\Pi(z_{(1)})) = \chi(\lambda)(z)$ for all $\lambda, \nu \in \mathfrak{h}^*$ and all Π'.

For $z_{(2)}$ in the second copy of the center, first suppose that $\nu \in \Lambda^+$. Write the Harish-Chandra projection of z, but now using the decomposition $\mathfrak{U}\mathfrak{g} = \mathfrak{U}\mathfrak{n}^+ \otimes \mathfrak{U}\mathfrak{h} \otimes \mathfrak{U}\mathfrak{n}^-$. In other words, compute $\beta^{w_\circ\Pi}(z)$, since $z_{(2)} \equiv (\beta^{w_\circ\Pi}(z))_{(2)}$ mod $(\mathfrak{U}\widehat{\mathfrak{g}})\widehat{\mathfrak{q}}$. Now use the basis $h_{i(2)}$ of $\mathfrak{h}_{(2)}$ to write out the above as a polynomial:

$$(\beta^{w_\circ\Pi}(z))_{(2)} = p(\{h_{i(2)} : i \in I\}) = p(\{\overline{h_i} - h_{i(1)} : i \in I\}).$$

Hence $\mathbf{h}^\Pi(z_{(2)}; \lambda, \nu)$ involves acting by ν on $\overline{h_i}$ and by λ on $h_{i(1)}$. Now recall that for all $z \in Z(\mathfrak{U}\mathfrak{g})$, $\overline{z} \in Z(\mathfrak{U}\overline{\mathfrak{g}}) \subset \Omega$. Using these facts, and fixing $\nu' \in \mathfrak{h}^*$, compute using Theorem 4.2:

$$\begin{aligned} \mathbf{h}^\Pi(z_{(2)}; \lambda, \nu) &= p(\{\nu(h_i) - \lambda(h_i)\}) = p(\{(\nu - \lambda)(h_i)\}) = (\nu - \lambda)(\beta^{w_\circ\Pi}(z))\nu'(1) \\ &= \mathbf{h}^{w_\circ\Pi}(\overline{z}; \nu', \nu - \lambda) = \mathbf{h}^\Pi(\overline{z}; w_\circ * \nu', w_\circ(\nu - \lambda)) = \chi(w_\circ(\nu - \lambda))(z), \end{aligned}$$

since $w_\circ = w_\circ^{-1}$. By Theorem 1.2, twist this weight by $w_\circ$; thus for all $(z, \lambda, \nu) \in Z(\mathfrak{U}\mathfrak{g}) \times \mathfrak{h}^* \times \Lambda^+$,

$$\mathbf{h}^\Pi(z_{(2)}; \lambda, \nu) = \chi(w_\circ * w_\circ(\nu - \lambda))(z) = \chi(\nu - \lambda - 2\rho)(z).$$

Now fix any weight λ and any central z, and consider the map $h_{z,\lambda} : \mathfrak{h}^* \to \mathbb{C}$, given by:

$$h_{z,\lambda}(\nu) := \mathbf{h}^\Pi(z_{(2)}; \lambda, \nu) - \chi(\nu - \lambda - 2\rho)(z).$$

It is clear that $h_{z,\lambda}$ is a polynomial map, which vanishes if ν lies in the Zariski dense subset $\Lambda^+ \subset \mathfrak{h}^*$. But then $h_{z,\lambda} \equiv 0$ as polynomials. (This is the aforementioned Zariski density argument; it is analogous to saying that if a polynomial $p(T_1, \ldots, T_n) : \mathbb{C}^n \to \mathbb{C}$ is identically zero on $z_0 + \mathbb{Z}_{\geq 0}^n$ for some $z_0 \in \mathbb{C}^n$, then $p \equiv 0$.) We conclude that for all $\lambda, \nu \in \mathfrak{h}^*$ and $z \in Z(\mathfrak{U}\mathfrak{g})$,

$\mathbf{h}^{\Pi}(z; \lambda, \nu) = \chi(\lambda, \nu - \lambda - 2\rho)(z)$. It is also easy to check that the following holds:

$$w * (\nu - \lambda - 2\rho) = w\nu - w * \lambda - 2\rho \ \forall w \in W, \nu, \lambda \in \mathfrak{h}^*. \tag{4.10}$$

Now given any $w \in W$, compute using Theorems 1.2 and 4.2:

$$\begin{aligned} \mathbf{h}^{w^{-1}\Pi}(z; \lambda, \nu) &= \mathbf{h}^{\Pi}(z; w * \lambda, w\nu) = \chi(w * \lambda, w\nu - w * \lambda - 2\rho)(z) \\ &= \chi(w * \lambda, w * (\nu - \lambda - 2\rho))(z) = \chi(\lambda, \nu - \lambda - 2\rho)(z). \end{aligned}$$

This proves the main assertion of the theorem; the rest follow easily. For instance, if $w\nu \in \Lambda^+$, then considering the action of any central z on the minimal type (and using Theorem 4.2),

$$\chi_{\widehat{\pi}_{\lambda,\nu}}(z) = \eta_{\widehat{\pi}_{\lambda,\nu}, \overline{\nu}}(z) = \mathbf{h}^{w^{-1}\Pi}(z; \lambda, \nu) = \chi(\lambda, \nu - \lambda - 2\rho)(z). \qquad \square$$

Remark 4.11. A similar result follows from the definitions: given $z \in Z(\mathfrak{Ug})$, $\lambda, \nu \in \mathfrak{h}^*$, and $w \in W$, $\mathbf{h}^{w^{-1}\Pi}(\overline{z}; \lambda, \nu) = \chi(w(\nu))(z)$. Note that this was shown in [PRV2] only when $w(\nu) \in \Lambda^+$. Now since the action of any $w \in W$ on $\mathfrak{h}^*$ is a linear - hence, polynomial - map, once again a Zariski density argument can be used to extend the result to all $\nu \in \mathfrak{h}^*$, for any fixed $z \in Z(\mathfrak{Ug})$.

In the above result, observe that the above recipe for the central character is "W-equivariant", in that the last equation in the statement of the theorem (which is basically Equation (4.10)) holds. One can check that this is not always so: in other words, if we use $\chi(\lambda, w * (\nu - \lambda - 2\rho))$ to denote the central character (via Theorem 1.2) for arbitrary $w \neq 1 \in W$ - such as $w = w_o$, say. However, one can check that W-equivariance does hold if w is central in W.

4.8. Remarks.

We conclude this section with some remarks. Note that the presentation in [PRV2] of the material in this section is motivated by the approach of Harish-Chandra. Thus, it differs somewhat from the presentation in this article.

More precisely, the philosophy in [PRV2] (as per the historical account given in [Va]) was to use Harish-Chandra's *density theorem*, which roughly says that among all irreducible G-modules containing a given irreducible finite-dimensional $\mathfrak{k}$-module $\mathcal{D}$, the ones that are finite-dimensional form a Zariski dense set. What this means is that the homomorphisms $\mathbf{h}^{\Pi'}(-; \lambda, \nu)$ correspond to finite-dimensional representations at special lattice points, as in Corollary 4.6 - and the set of these lattice points is Zariski dense in $\mathfrak{h}^* \times \mathfrak{h}^*$. If we replace these by more general points, then the homomorphisms in question correspond to infinite-dimensional representations.

In view of this perspective, the approach in [PRV2] was to "alternate" between the representations and the homomorphisms. Here is their strategy:

- Identify and study the minimal type in all finite-dimensional simple $\widehat{\mathfrak{g}}$-modules $V(\lambda, \mu)$.

- Explicitly compute the polynomials $\mathbf{h}^{\Pi'}$ for these modules, and prove that $\eta_{V(\lambda,\mu),\overline{\nu}}(-) \equiv \mathbf{h}^{w^{-1}\Pi}(-; \lambda, \nu)$, where $\nu = \lambda + w_\circ\mu$ and $\overline{\nu} = w(\nu) \in \Lambda^+$.

- Motivated by this, claim that there are simple modules $\widehat{\pi}_{\lambda,\nu}$ in $\mathcal{C}(\mathfrak{g} \times \mathfrak{g}, \overline{\mathfrak{g}})$ for all $\lambda \in \mathfrak{h}^*$ and $\nu \in \Lambda$, with minimal type $\overline{\nu}$; now construct these.

- Finally, prove that $\eta_{\widehat{\pi}_{\lambda,\nu},\overline{\nu}}(-)$ indeed equals $\mathbf{h}^{w^{-1}\Pi}(-; \lambda, \nu)$ for these modules.

5. The Rings $\mathcal{R}_{\nu,\Pi'}$ and the PRV Determinants

The next object of study in [PRV2] is the image $\mathcal{R}_{\nu,w^{-1}\Pi}$ of the map $\mathbf{h}^{w^{-1}\Pi}(-; -, \nu) : \Omega \to \operatorname{Sym}\mathfrak{h}$, where $w(\nu) \in \Lambda^+$. This was done in detail in [PRV2, Section 3] using deep results of Kostant [Ko2], such as his separation of variables theorem (2.4). These results lead to the definition and study of the so-called PRV determinants, which we discuss below.

5.1. The rings $\mathcal{R}_{\nu,\Pi'}$. To state the next result, we need some notation. Recall the stabilizer subgroup W_ν of a weight ν, as well as the twisted action of the Weyl group $*$ on $\mathfrak{h}^*$, which transfers to $\mathfrak{h}$ and then extends to $\operatorname{Sym}\mathfrak{h} = P(\mathfrak{h}^*)$.

Theorem 5.1 ([PRV2]). *For all $\nu \in \Lambda$ and $w \in W$ such that $w(\nu) \in \Lambda^+$, $\mathcal{R}_{\nu,w^{-1}\Pi} \subset \mathbf{I}_\nu$, where*

$$\mathbf{I}_\nu := \{p \in P(\mathfrak{h}^*) : w * p = p \; \forall w \in W_\nu\}.$$

If $\nu = 0$ or $\nu(h_i) > 0$ for all $i \in I$, then $\mathcal{R}_{\nu,\Pi} = \mathbf{I}_\nu$. For general $\nu \in \Lambda^+$, define

$$(\mathfrak{U}\mathfrak{g})_0(\nu) := \{a \in (\mathfrak{U}\mathfrak{g})_0 : (\operatorname{ad} e_i)^{\nu(h_i)+1}(a) = 0 \; \forall i \in I\}.$$

Then $\mathcal{R}_{\nu,\Pi} = w_\circ * \beta^\Pi((\mathfrak{U}\mathfrak{g})_0(-w_\circ\nu))$, and $\mathcal{R}_{w\nu,w\Pi} = w * \mathcal{R}_{\nu,\Pi}$ for all $w \in W$.

Note that if $\nu(h_i) > 0$ for all i (i.e., ν is regular), then $\mathbf{I}_\nu = P(\mathfrak{h}^*)$. Moreover, $\mathbf{I}_0 = P(\mathfrak{h}^*)^{(W,*)}$.

Remark 5.2. The authors claim in [PRV2] that they have proved that $\mathcal{R}_{\nu,\Pi} = \mathbf{I}_\nu$ for all $\nu \in \Lambda^+$, using a case-by-case analysis.

114

We also remark that Theorem 5.1 has the following consequence for $\nu = 0$ or regular ν.

Corollary 5.3 ([PRV2]). *Suppose* $\nu(h_i) > 0$ *for all* i, *and* $\lambda, \lambda' \in \mathfrak{h}^*$. *Then,*

$$\widehat{\pi}_{\lambda,0} \cong \widehat{\pi}_{\lambda',0} \Longleftrightarrow \lambda' \in W * \lambda,$$

whereas the representations $\widehat{\pi}_{\lambda,\nu}$ *are inequivalent for all* λ.

The proofs of these results are carefully developed in [PRV2, Section 3], via many intermediate lemmas. These lemmas heavily use results developed by Kostant in [Ko2], which deal with the symmetrization map and with finite-dimensional $\mathfrak{g}$-submodules of Sym $\mathfrak{g}$ and $\mathfrak{Ug}$. In the next part, we discuss some of these results, and show how they can be used to define and study the "PRV determinants".

We end this part by discussing the case of $\mathfrak{sl}_2$, where we classify all of the modules $\widehat{\pi}_{\lambda,\nu}$.

Example 5.4. Suppose $\mathfrak{g} = \mathfrak{sl}_2(\mathbb{C})$. Using the results of Section 4.6, compute with $\nu \in \mathbb{Z}_{\geq 0}$:

$$\mathbf{h}^{\Pi}(\Delta_1; -, \nu) = h_1^2 + 2h_1, \quad \mathbf{h}^{\Pi}(\Delta_2; -, \nu) = \nu^2 - 2\nu - 2\nu h_1 + h_1^2 + 2h_1, \quad \mathbf{h}^{\Pi}(\overline{\Delta}; -, \nu) = \nu^2 + 2\nu.$$

It is now clear that if $\nu = 0$, then these generate $\mathbb{C}[h_1^2 + 2h_1] = \mathbb{C}[h_1]^{(W,*)}$. The above result says that $\mathbf{h}^{\Pi}(\omega; -, \nu)$ is an element of this ring for all $\omega \in \Omega$. Similarly, if $\nu > 0$ (abusing notation), then the above polynomials already generate all of $\mathbb{C}[h_1]$ as desired.

Also note that we can classify (all equivalences between) the representations $\widehat{\pi}_{\lambda,\nu}$: given $(\lambda, \nu) \neq (\lambda', \nu')$ in $\mathfrak{h}^* \times \Lambda$, we claim:

$$\widehat{\pi}_{\lambda,\nu} \cong \widehat{\pi}_{\lambda',\nu'} \Longleftrightarrow (\lambda', \nu') = (-\lambda - 2, -\nu). \tag{5.5}$$

The backward implication follows from Corollary 4.3, and conversely, $\nu' = \pm\nu$. Then the calculations above imply that $\mathbf{h}^{\Pi}(\Delta_1; \lambda, \nu) = \mathbf{h}^{\Pi}(\Delta_1; \lambda', \nu')$, whence $\lambda' = \lambda, -\lambda - 2$. This shows the claim when $\nu' = \pm\nu = 0$. Otherwise we may assume that $\nu' = \nu > 0$ (using the backward implication). Now evaluate $\mathbf{h}^{\Pi}(\overline{\Delta} + \Delta_1 - \Delta_2; -, \nu)$ at λ, λ'. Then $\nu \neq 0 \implies \lambda = \lambda'$. $\square$

These calculations naturally lead to the question of classifying the representations $\widehat{\pi}_{\lambda,\nu}$ for general semisimple $\mathfrak{g}$ - and more generally, the classification of all simple objects of $\mathcal{C}(\mathfrak{g} \times \mathfrak{g}, \overline{\mathfrak{g}})$. (Recall Corollary 4.3.) These questions will be discussed in the concluding section of this article.

5.2. Kostant's separation of variables.

In order to discuss PRV determinants, we need some preliminaries. Recall the symmetrization map $\boldsymbol{\lambda}$ from [Har2], which is the unique linear isomorphism : $\operatorname{Sym} \mathfrak{g} \twoheadrightarrow \mathfrak{U}\mathfrak{g}$ satisfying: $\boldsymbol{\lambda}(1) = 1$ and for all $r > 0$ and $X_1, \ldots, X_r \in \mathfrak{g}$,

$$\boldsymbol{\lambda}(X_1 \ldots X_r) = \frac{1}{r!} \sum_{\sigma \in S_r} X_{\sigma(1)} \ldots X_{\sigma(r)}.$$

Also recall that the adjoint action of $\mathfrak{g}$ on itself can be uniquely extended to derivations $\Phi : \mathfrak{g} \to \operatorname{Sym} \mathfrak{g}$ and $\Theta : \mathfrak{g} \to \mathfrak{U}\mathfrak{g}$. Thus, both of these algebras are $\mathfrak{g}$-modules, and $\boldsymbol{\lambda}$ is a $\mathfrak{g}$-module isomorphism: $\boldsymbol{\lambda} \circ \Phi(X) = \Theta(X) \circ \boldsymbol{\lambda}$ for all $X \in \mathfrak{g}$. Moreover, the centers are isomorphic via $\boldsymbol{\lambda}$. Namely, $\boldsymbol{\lambda} : (\operatorname{Sym} \mathfrak{g})^{G_0} \twoheadrightarrow Z(\mathfrak{U}\mathfrak{g}) = (\mathfrak{U}\mathfrak{g})^{G_0}$, where G_0 is the adjoint group of $\mathfrak{g}$.

Now given $\mu \in \Lambda^+$, define the following copies of the $\mathfrak{g}$-module $V(\mu)$ in $\operatorname{Sym} \mathfrak{g}$ and $\mathfrak{U}\mathfrak{g}$:

$$\mathcal{L}^\mu := \operatorname{Hom}_{\mathfrak{g}}(V(\mu), \operatorname{Sym} \mathfrak{g}), \quad \mathcal{L}^\mu(\mathfrak{U}\mathfrak{g}) := \operatorname{Hom}_{\mathfrak{g}}(V(\mu), \mathfrak{U}\mathfrak{g}) = \{\boldsymbol{\lambda} \circ M : M \in \mathcal{L}^\mu\}.$$

In [Ko2], Kostant showed that these are both free modules of rank $d_\mu := \dim V(\mu)_0$, over $(\operatorname{Sym} \mathfrak{g})^{G_0}$ and $Z(\mathfrak{U}\mathfrak{g})$ respectively. This is related to Kostant's "separation of variables" theorem; see Equation (2.4), where $\mathbb{H}(\mathfrak{g})$ is precisely the image of the harmonic polynomials in $\operatorname{Sym} \mathfrak{g}$ under $\boldsymbol{\lambda}$. (Recall from above that $[\mathbb{H}(\mathfrak{g}) : V(\mu)] = \dim V(\mu)_0 = d_\mu$.) Moreover, it is possible to choose all of the basis elements M_i (over the center) for $\mathcal{L}^\mu$ such that every $M_i(V(\mu))$ is a subspace of $\operatorname{Sym}^{q_i(\mu)} \mathfrak{h}$ for some $q_i(\mu) \in \mathbb{Z}_{\geq 0}$. Such elements are called *homogeneous*. The image of a set of homogeneous generators under $\boldsymbol{\lambda}$ yields a set of $Z(\mathfrak{U}\mathfrak{g})$-module generators for $\mathcal{L}^\mu(\mathfrak{U}\mathfrak{g})$.

The zero weight spaces in these modules $\mathcal{L}^\mu, \mathcal{L}^\mu(\mathfrak{U}\mathfrak{g})$ are of great interest in [PRV2]. For instance, it is easy to check that for all $\nu \in \Lambda^+$,

$$(\mathfrak{U}\mathfrak{g})_0(\nu) = \sum_{\mu \in \Lambda^+} \sum_{L \in \mathcal{L}^\mu(\mathfrak{U}\mathfrak{g})} L(V^+(\mu; 0, \nu)),$$

where $V^+(\mu; 0, \nu)$ was defined above Theorem 3.1. (These spaces were used in Theorem 5.1.)

5.3. PRV determinants.

We finally define the *PRV determinants*. (This material is taken from [PRV2, Sections 3,4].) Fix $\mu \in \Lambda^+$, and choose a set $\{M_1, \ldots, M_{d_\mu}\}$ of homogeneous generators for $\mathcal{L}^\mu$. Also choose a basis $\{v_1, \ldots, v_{d_\mu}\}$ of $V(\mu)_0$. Now define the *PRV matrix* as:

$$\mathbf{K}'_\mu := ((\beta^\Pi(\boldsymbol{\lambda}(M_i)v_j)))_{1 \leq i,j \leq d_\mu} \in \mathfrak{gl}_{d_\mu}(\operatorname{Sym} \mathfrak{h}), \quad \mathbf{K}'_\mu(\nu) = ((\nu((\mathbf{K}'_\mu)_{ij}))) \in \mathfrak{gl}_{d_\mu}(\mathbb{C}).$$

One of the important results in [PRV2] that has also influenced much future research is the computation of the determinants of these and related matrices. In the following preliminary result, recall by Theorem 3.1 that $\dim V^+(\mu; 0, -w_\circ\nu) = m^\mu_{-w_\circ\nu,\nu} = [V_{\bar{\mathfrak{g}}}(\nu) \otimes V_{\bar{\mathfrak{g}}}(\nu)^* : V_{\bar{\mathfrak{g}}}(\mu)]$.

Proposition 5.6 ([PRV2]). *For all* $\mu, \nu \in \Lambda^+$, *rank* $\mathbf{K}'_\mu(\nu) = \dim V^+(\mu; 0, -w_\circ\nu)$. *Moreover,*

$$d_\mu > 0, \lambda \in \mathfrak{h}^* \implies [\hat{\pi}_{\lambda,0} : V_{\bar{\mathfrak{g}}}(\mu)] \leq \min(\operatorname{rank} \mathbf{K}'_{-w_\circ\mu}(\lambda), \operatorname{rank} \mathbf{K}'_\mu(w_\circ * \lambda)).$$

The proposition is proved using the auxiliary lemmas developed in [PRV2, Section 3], and holds for all dominant integral μ, ν. The authors now propose another matrix which turns out to be nonsingular at all regular points $h \in \mathfrak{h}$. To define this, suppose $M_i \in \mathcal{L}^\mu$, $v_j \in V(\mu)_0$, and $q_i(\mu) = \deg(M_i v_j) \geq 0$ as above. Now there exists a unique $h_{ij} \in \operatorname{Sym}\mathfrak{h}$ such that $M_i v_j \equiv h_{ij}$ mod $\sum_{\alpha \in R}(\operatorname{Sym}\mathfrak{h})\mathfrak{g}_\alpha$, and moreover, h_{ij} is homogeneous of degree $q_i(\mu)$ for all $1 \leq i, j \leq d_\mu$. We now introduce the following terminology.

Definition 5.7.

1. Define the matrix $\mathbf{K}_\mu := ((h_{ij}))_{1 \leq i,j \leq d_\mu}$.

2. Fix $\alpha \in R^+$, and suppose $\alpha \leftrightarrow h'_\alpha$ via the Killing form. Now define $h_\alpha := (2/\alpha(h'_\alpha))h'_\alpha$. Let e_α, f_α span $\mathfrak{g}_\alpha, \mathfrak{g}_{-\alpha}$ respectively, and let $m_{j,\mu}(\alpha)$ denote the multiplicity of the eigenvalue $j(j+1)$ for the restriction of $f_\alpha e_\alpha$ to $V(\mu)_0$. Finally, define $m_\mu(\alpha) = \sum_{j>0} m_{j,\mu}(\alpha)$.

The following is the main theorem of [PRV2] involving PRV determinants:

Theorem 5.8 ([PRV2]). *Fix* $\mu \in \Lambda^+$. *Viewed (via the Killing form) as a polynomial function on* $\mathfrak{h}$, *det* $\mathbf{K}_\mu$ *is nonzero at each regular point* $h \in \mathfrak{h}$. *In particular,* $\det \mathbf{K}_\mu, \det \mathbf{K}'_\mu$ *are nonzero elements of* $P(\mathfrak{h}^*)$ *of degree* $\sum_i q_i(\mu)$. *Moreover, there exist nonzero constants* c_μ, c'_μ *such that:*

$$\det \mathbf{K}_\mu = c_\mu \prod_{\alpha \in R^+} h_\alpha^{m_\mu(\alpha)}, \qquad \det \mathbf{K}'_\mu = c'_\mu \prod_{\alpha \in R^+} \prod_{j \geq 1} \{h_\alpha + \rho(h_\alpha) - 1, j\}^{m_{j,\mu}(\alpha)},$$

where for all $a \in \mathfrak{Ug}$ *and* $j \in \mathbb{N}$, $\{a, j\} := (-1)^j j! \, a(a-1)\ldots(a-j+1)$. *(In particular,* $\sum_{i=1}^{d_\mu} q_i(\mu) = \sum_{\alpha \in R^+} m_\mu(\alpha)$.)

This is a powerful theorem that computes the PRV determinants in a simple manner. It can be used to compute these determinants explicitly for simple $\mathfrak{g}$, with $V(\mu)$ the adjoint representation, for instance. Let us take a simple example: $\mathfrak{g} = \mathfrak{sl}_2$. First note that $d_\mu > 0$ if and only if $\mu \in 2\mathbb{Z}_{\geq 0}\varpi$. Suppose

117

this holds, and apply Equation (1.1). Then, $fe \cdot v_0 = (\mu/2)(\mu/2 + 1)v_0$, so $m_{j,\mu}(\alpha) = \delta_{j,\mu/2}$ and $m_\mu(\alpha) = 1$ for all even μ. Hence

$$\det \mathbf{K}_\mu = c_\mu h, \qquad \det \mathbf{K}'_\mu = c'_\mu \{h, \mu/2\} \in \mathbb{C}^\times \cdot h(h-1)\dots(h-\mu/2+1).$$

It turns out that this is exactly the Shapovalov determinant for $\mathfrak{sl}_2(\mathbb{C})$ (up to a scalar). We now show how these are related to PRV determinants, before moving on to the next section.

5.4. Annihilators of Verma modules.

Since they were defined and computed in [PRV2], PRV determinants have played a role in the study of annihilators of Verma modules and their simple quotients, in the following manner. In [Du1], Duflo proved the following remarkable result: *The annihilator of every Verma module is centrally generated.* In other words, for any complex semisimple $\mathfrak{g}$,

$$\mathrm{Ann}_{\mathfrak{Ug}}\, M(\lambda) = \mathfrak{Ug} \cdot \mathrm{Ann}_{Z(\mathfrak{Ug})}\, M(\lambda) = \mathfrak{Ug} \cdot \ker \chi(\lambda), \qquad \forall \lambda \in \mathfrak{h}^*. \quad (5.9)$$

The proof of this statement requires a nontrivial algebro-geometric argument from [Ko2]. Thus, it cannot be extended to the setting of quantum groups $U_q(\mathfrak{g})$, and a new proof was sought. This was provided by Joseph for quantum groups, but it holds in the classical setting as well. In this part, we discuss how PRV determinants play a role in proving Duflo's result.

Given a weight $\mu \in \mathfrak{h}^*$, every weight space $M(\mu)_{\mu-\beta}$ of a Verma module has a bilinear form defined on it (where $\beta \in \mathbb{Z}_{\geq 0}\Pi$). To see this, first fix a Lie algebra anti-involution $\iota : \mathfrak{g} \to \mathfrak{g}$ that fixes $\mathfrak{h}$ and sends e_i to f_i for all $i \in I$. Then ι sends $\mathfrak{g}_\alpha$ to $\mathfrak{g}_{-\alpha}$ for all roots α, and also extends to an algebra anti-involution of $\mathfrak{Ug}$. Now the *Shapovalov form* is defined as follows:

$$\mathrm{Sh}(b_1, b_2) := \beta^\Pi(\iota(b_1) \cdot b_2) \in \mathrm{Sym}\,\mathfrak{h}, \quad \mathrm{Sh}_\mu(b_1 m_\mu, b_2 m_\mu) := \mu(\mathrm{Sh}(b_1, b_2)), \quad \forall b_1, b_2 \in \mathfrak{Un}^-,$$

where $m_\mu \in M(\mu)_\mu$ generates the Verma module. One shows that the nondegeneracy of this form can be checked on each individual weight space. Thus for all $\nu \in \mathbb{Z}_{\geq 0}\Pi$, let $\det \mathrm{Sh}^\nu$ be the determinant of Sh, when restricted to a (fixed) weight space basis of $(\mathfrak{Un}^-)_{-\nu}$. The Shapovalov form can now be shown to possess the following properties (see [MP, Sh] for instance):

Theorem 5.10. *For all $\mu \in \mathfrak{h}^*$, Sh_μ is a symmetric bilinear form on $M(\mu)$. $\mathrm{Sh}_\mu(M(\mu)_\nu, M(\mu)_{\nu'})$ is nonzero only when $\nu = \nu' \in \mu - \mathbb{Z}_{\geq 0}\Pi$. Moreover, the radical of Sh_μ is the unique maximal submodule of $M(\mu)$, so it is nondegenerate if and only if $M(\mu) \cong V(\mu)$. Finally, there exists a nonzero constant c''_ν such that*

$$\det \mathrm{Sh}^\nu = c''_\nu \prod_{\alpha \in R^+} \prod_{j \geq 1} (h_\alpha + \rho(h_\alpha) - j)^{\mathcal{P}(\nu - j\alpha)}.$$

We note that both the PRV and Shapovalov determinants $\det \mathbf{K}'_\mu, \det \mathrm{Sh}^\nu$ are products of linear factors. However, more holds! (The rest of this part is from [FL].)

Theorem 5.11. *For all semisimple $\mathfrak{g}$, the set of all linear factors in $\{\det \mathbf{K}'_\mu : \mu \in \Lambda^+, d_\mu > 0\}$ and $\{\det \mathrm{Sh}^\nu : \nu \in \mathbb{Z}_{\geq 0}\Pi\}$ coincide. Moreover, given $\lambda \in \mathfrak{h}^*$, the annihilator $\mathrm{Ann}_{\mathbb{H}(\mathfrak{g})} V(\lambda)$ is trivial if and only $\lambda(\det \mathbf{K}'_\mu) = 0$ whenever $d_\mu > 0$.*

Thus, another approach to proving Duflo's "Verma module annihilator theorem" (5.9) is to proceed as follows. This is a program developed by Joseph and his coauthors.

1. $U := \mathfrak{Ug}$ has a large "locally finite subalgebra" $F(U) := \{a \in U : \dim(\mathrm{ad}\, U)a < \infty\}$.

2. A "Peter-Weyl" type result holds: $F(U) := \bigoplus_{\lambda \in \Lambda^+} \mathrm{End}_\mathbb{C} V(\lambda)$. (Note that for $U = \mathfrak{Ug}$, this is proved for $F(U) = U$ using the perfect pairing between left-invariant differential operators in $\mathfrak{Ug}$ and regular functions on the simply connected Lie group G for which $\mathfrak{g} = \mathrm{Lie}(G)$, together with the usual Peter-Weyl Theorem for regular functions on G.) Under this identification, the lifts of the identity elements in the various summands are a basis $\{z_\lambda : \lambda \in \Lambda^+\}$ of the center; moreover, $Z(U)$ is a polynomial algebra in the generators $\{z_{\varpi_i} : i \in I\}$.

3. There exists an $\mathrm{ad}\, U$-stable submodule $\mathbb{H}$ of $F(U)$ such that the multiplication map : $\mathbb{H} \otimes Z(U) \to F(U)$ is an isomorphism of $\mathrm{ad}\, U$-modules.[6] Here, ad is the standard adjoint action of the Hopf algebra U on itself.

4. Now define the PRV determinants using the above facts, and the Shapovalov determinants using the anti-involution ι and the Harish-Chandra projection β^Π. Then calculate both sets of determinants and verify their properties as in the above results.

5. Now use the PRV and Shapovalov determinants for any simple submodule of a Verma module $M(\lambda)$, which is itself a Verma module with the same central character $\chi(\lambda)$. Some more work now shows Duflo's result.

The important point is that this approach works not only for $U = \mathfrak{Ug}$, but also for $U = U_q(\mathfrak{g})$. (Unlike $\mathfrak{Ug}$, $F(U) \neq U$ in the quantum case.) Thus, Joseph and Letzter proved the quantum "separation of variables" theorem, and defined

---

[6]If $U = \mathfrak{Ug}$, then $F(U) = U$, and this result is precisely Equation (2.4). One can also use this and Example 3.2 to prove the Peter-Weyl type result mentioned above, by counting (countably infinite) multiplicities.

and computed the related PRV determinants. See [JL1, JL2]; also see [FL] for a historical exposition of this program.

Joseph and others have since extended this approach to affine Lie algebras. Similarly, Gorelik and Lanzmann [GL] have also carried out this program for reductive super Lie algebras. They found that the PRV determinants contained some "extra factors" compared to the Shapovalov determinants, and their zeroes are precisely the weights for which the corresponding Verma module annihilators are not centrally generated. Thus, the PRV and Shapovalov determinants (or more precisely, their common zeroes) turn out to yield, in various settings, both an approach to proving Duflo's Theorem (5.9), as well as the set of Verma modules for which it holds.

5.5. KPRV determinants. We end this section with a remark. Kostant described certain analogues of the PRV determinants in [Ko3] involving parabolic subalgebras of $\mathfrak{g}$; these analogues had applications related to the irreducibility of principal series representations. Joseph termed these the *KPRV determinants*, and together with Letzter and with Todoric, has defined such notions for (quantum) semisimple and affine Lie algebras. (See [Jo3, JL3, JLT, JT] for more on this, including applications to annihilators of Verma modules.) Thus, the PRV determinants and their generalizations continue to be a useful and popular subject of research in several different settings in representation theory.

6. Representations of Class Zero

In [PRV2, Section 4], the authors apply the theory previously developed to carry out a deeper study of a special sub-family of irreducible admissible $(\mathfrak{g} \times \mathfrak{g}, \bar{\mathfrak{g}})$-modules: the ones of "class zero". Here is a brief discussion of these modules and related results.

Definition 6.1. An irreducible $\bar{\mathfrak{g}}$-admissible G-representation V is said to be of *class zero* if $[V : V_{\bar{\mathfrak{g}}}(0)] > 0$.

(Note that these are usually referred to as "class one" representations in the literature - i.e., irreducible admissible (G, K)-modules which have a K-invariant vector. In other words, "class one" refers to a K-eigenvector with simultaneous eigenvalue 1, while "class zero" refers to a $\bar{\mathfrak{g}}$-eigenvector with simultaneous eigenvalue 0.) The first result says that every irreducible Harish-Chandra module of class zero is determined by its central character; in fact, it is of the form $\hat{\pi}_{\lambda,0}$ for some $\lambda \in \mathfrak{h}^*$. Thus, we obtain deeper insights into the classification of such modules. (See Equation (2.8) and Theorem 4.9.)

120

Theorem 6.2 ([PRV2]). *The set of infinitesimal equivalence classes of class zero irreducible representations V is in bijection with the twisted Weyl group orbits in $\mathfrak{h}^*$. More precisely, every such V is uniquely determined by its infinitesimal character χ_V restricted to $Z(\mathfrak{U}\mathfrak{g})_{(1)}$. Moreover, given $\chi_V : Z(\mathfrak{U}(\mathfrak{g} \times \mathfrak{g})) \to \mathbb{C}$, there exists $\lambda \in \mathfrak{h}^*$ such that*

$$\chi_V = \chi(\lambda, -\lambda - 2\rho), \qquad V \cong \widehat{\pi}_{\lambda,0} \cong \widehat{\pi}_{w*\lambda,0} \ \forall w \in W.$$

In particular $[V : V_{\overline{\mathfrak{g}}}(0)] = 1$ and 0 is the minimal type of V.

Example 6.3. If $\lambda \in \Lambda^+$, then $\chi_{\widehat{\pi}_{\lambda,0}} = \chi(\lambda, w_\circ * (-\lambda - 2\rho)) = \chi(\lambda, -w_\circ \lambda) = \chi_{V(\lambda) \otimes V(\lambda)^*}$. By Theorem 3.5, the minimal type of $V(\lambda) \otimes V(\lambda)^*$ is $\lambda + w_\circ(-w_\circ \lambda) = 0$ as well. Therefore $\widehat{\pi}_{\lambda,0} \cong V(\lambda) \otimes V(\lambda)^*$ by the above result. Moreover, every finite-dimensional $\widehat{\pi}_{\lambda,0}$ is of this kind, by Corollary 4.6.

We now take a closer look at the multiplicities. When is $[\widehat{\pi}_{\lambda,0} : V_{\overline{\mathfrak{g}}}(\mu)] = \dim V_{\overline{\mathfrak{g}}}(\mu)_0$? (That it is at most $\dim V_{\overline{\mathfrak{g}}}(\mu)_0$ follows from Theorem 4.1.) More generally, is it possible to compute the multiplicities for class zero modules? Once again, the authors were able to achieve this goal in [PRV2]: the multiplicity equals the rank of a related matrix, which is defined similar to $\mathbf{K}'_\mu$ above. Namely, given $\mu \in \Lambda^+$ such that $d_\mu > 0$, choose sets of homogeneous generators $\{M_1, \ldots, M_{d_\mu}\}$ and $\{M_1^*, \ldots, M_{d_\mu}^*\}$ for the free $(\text{Sym }\mathfrak{g})^{G_0}$-modules $\mathcal{L}^\mu$ and $\mathcal{L}^{-w_\circ \mu}$ respectively. Now choose dual bases $\{v_k\}$ and $\{v_k^*\}$ for $V(\mu), V(-w_\circ \mu) \cong V(\mu)^*$ respectively. The span of

$$z_{ij} := \sum_k (\boldsymbol{\lambda}(M_i^*)v_k^*)(\boldsymbol{\lambda}(M_j)v_k)$$

then depends only on M_i and M_j. One can now compute the sought-for multiplicities.

Theorem 6.4 ([PRV2]). *For all $\mu \in \Lambda^+$ and $1 \leq i, j \leq d_\mu$, $z_{ij} \in Z(\mathfrak{U}\mathfrak{g})$. Now given a central character $\chi = \chi(\lambda)$ of $Z(\mathfrak{U}\mathfrak{g})$ (where $\lambda \in \mathfrak{h}^*$), $[\widehat{\pi}_{\lambda,0} : V_{\overline{\mathfrak{g}}}(\mu)] = \mathbf{1}_{d_\mu > 0} \cdot \text{rank}((\chi(\lambda)(z_{ij})))$.*

The next result discusses the case when the multiplicities all attain their upper bounds. This turns out to be an important question from the point of view of the irreducibility of the induced representations $\pi_{\xi,\nu}$ discussed earlier; see also [Bru]. The following result completely answers this question. (See [Du2] for more results along these lines.)

Theorem 6.5 ([PRV2]). *Given $\lambda \in \mathfrak{h}^*$, the following are equivalent:*

1. $\widehat{\pi}_{\lambda,0}$ *is complete, i.e., $[\widehat{\pi}_{\lambda,0} : V_{\overline{\mathfrak{g}}}(\mu)] = \dim V_{\overline{\mathfrak{g}}}(\mu)_0$ whenever $d_\mu > 0$ for $\mu \in \Lambda^+$.*

2. $\pi_{2(\lambda+\rho),0}$ *is irreducible.*

3. *The matrices* $\mathbf{K}'_{-w_\circ \mu}(\lambda)$ *and* $\mathbf{K}'_\mu(w_\circ * \lambda)$ *are both invertible whenever* $d_\mu > 0$ *for* $\mu \in \Lambda^+$.

4. *For all roots* $\alpha \in R^+$, $\frac{1}{2}\xi(h_\alpha) = (\lambda + \rho)(h_\alpha) \notin \mathbb{Z} \setminus \{0\}$.

This result now holds when ξ attains purely imaginary values on $\mathfrak{h}_0$ (i.e., $\xi \in \mathfrak{R} \setminus \{0\}$, in the notation of Theorem 4.1). This shows the irreducibility of a class of unitary representations that was studied previously in the complex semisimple case:

Theorem 6.6 ([PRV2]). *The unitary G-representations of the principal non-degenerate series (of Gelfand and Naimark [GN]) that contain a nonzero K-invariant vector, are all irreducible.*

7. Conclusion: The Classification of Irreducible Harish-Chandra Modules

As discussed in previous sections, the paper [PRV2] has led to much research in several different directions in representation theory. In this final section, we return to its original motivations. As is evident from the paper, as well as from much of the contemporary literature, the fundamental and profound work of Harish-Chandra on semisimple (real) Lie groups has had an enormous influence on the field of representation theory. From the objects studied to the methods employed in [PRV2], the authors have time and again used contributions of Harish-Chandra to the subject.

We now mention some of the subsequent developments in the program started by Harish-Chandra, of studying K-admissible G-representations. For instance, various results from [Har2, Har3, PRV2] were subsequently generalized by Lepowsky in [Le]. Moreover, in [Zh1, Zh2], Zhelobenko classified irreducible admissible representations of complex semisimple Lie groups, by showing that they always arise as distinguished quotients of certain principal series representations. This classification is the subject of this section.

Before moving on to these classification results, we remark that this program was extended by Langlands in [La], to the original setting of real semisimple Lie groups $G_\mathbb{R}$, where Harish-Chandra had introduced and studied admissible representations. The *Langlands classification* describes how irreducible admissible representations are quotients of "generalized principal series", which are induced from tempered representations on parabolic subgroups of $G_\mathbb{R}$. The work of Langlands and Harish-Chandra on tempered representations was refined by Knapp and Zuckerman [KZ]; thus, one now has an explicit parametrization of the irreducible admissible representations of the groups $G_{\mathbb{R}}$. (See also [BB], which studies more general irreducible representations than just the admissible ones.) As this suggests, the legacy of Harish-Chandra is vast and rich, and lives on in these works and in the subsequent research which it has inspired.

7.1. The set of irreducible objects.

We now discuss the classification of all irreducible objects of $\mathcal{C}(\mathfrak{g} \times \mathfrak{g}, \bar{\mathfrak{g}})$. There are two parts to this discussion: first, to determine a representative set of simple objects that covers all isomorphism classes; and second, to determine the equivalences among the objects in this set. In what follows, the final results will be stated as they appear in Duflo's notes [Du2] on the subject.

It turns out that the category $\mathcal{C}(\mathfrak{g} \times \mathfrak{g}, \bar{\mathfrak{g}})$ is equivalent to a subcategory of the BGG Category $\mathcal{O}$. In particular, Harish-Chandra modules have certain properties in common with Verma modules. For instance, all objects of this category have finite length, all simple objects have a corresponding central character, and for a given central character, the simple objects are indexed by the Weyl group. More precisely, Beilinson and Bernstein have classified all irreducible $(\mathfrak{g} \times \mathfrak{g}, \bar{\mathfrak{g}})$-modules with a fixed infinitesimal character. Here is a special case of their results.

Theorem 7.1 ([BB]). *Given $\lambda, \mu \in \Lambda^+$, the set of isomorphism classes of irreducible admissible $(\mathfrak{g} \times \mathfrak{g}, \bar{\mathfrak{g}})$-modules with infinitesimal character $\chi = \chi(\lambda, \mu)$ is in bijection with $W_\lambda \backslash W / W_\mu$.*

For instance, if λ, μ are both regular, then there are exactly $|W|$ isomorphism classes, while there is a single class if λ or μ is zero.

Now recall Theorem 6.2, which says that all irreducible admissible class zero modules are of the form $\widehat{\pi}_{\lambda,0}$ for some $\lambda \in \mathfrak{h}^*$. One can similarly ask: is every irreducible admissible $(\mathfrak{g} \times \mathfrak{g}, \bar{\mathfrak{g}})$-module of the form $\widehat{\pi}_{\lambda,\nu}$ for some $(\lambda, \nu) \in \mathfrak{h}^* \times \Lambda$? The answer turns out to be positive.

Theorem 7.2. *Suppose V is an irreducible object of $\mathcal{C}(\mathfrak{g} \times \mathfrak{g}, \bar{\mathfrak{g}})$. Then there exist $\lambda \in \mathfrak{h}^*$ and $\nu \in \Lambda$ such that $V \cong \widehat{\pi}_{\lambda,\nu} \cong \widehat{\pi}_{w*\lambda,w\nu}$ for all $w \in W$. In particular, V has minimal type $\bar{\nu}$ and infinitesimal character $\chi(\lambda, \nu - \lambda - 2\rho)$.*

7.2. The objects in a given isomorphism class.

The other aspect of classification is to identify the isomorphism classes. In light of the above result, the task is to identify when $\widehat{\pi}_{\lambda,\nu} \cong \widehat{\pi}_{\lambda',\nu'}$. In light of Corollary 4.3, one may assume that $\nu = \nu' \in \Lambda^+$. Moreover, in light of Corollaries 4.3 and 5.3 for general $\mathfrak{g}$, and Equation (5.5) for $\mathfrak{sl}_2(\mathbb{C})$, it is easy to guess the general result. This is further reinforced by the fact that if $\nu, \nu' \in \Lambda$, and $\xi \in \mathfrak{R}_\nu, \xi' \in \mathfrak{R}_{\nu'}$ (notation as in Theorem 4.1), then $\pi_{\xi,\nu}$ is irreducible, hence isomorphic to $\widehat{\pi}_{\lambda,\nu}$

123

(and similarly for $\pi_{\xi',\nu'}$). Now as observed in [PRV2],

$$\widehat{\pi}_{\lambda,\nu} \cong \widehat{\pi}_{\lambda',\nu'} \iff \pi_{\xi,\nu} \cong \pi_{\xi',\nu'} \iff \Theta_{\xi,\nu} = \Theta_{\xi',\nu'}$$
$$\iff \exists w \in W : (\xi',\nu') = (w\xi, w\nu) \iff \exists w \in W : (\lambda',\nu') = (w * \lambda, w\nu).$$

It should not come as a surprise now, that the obvious claim turns out to be correct:

Theorem 7.3. *Given* $(\lambda,\nu), (\lambda',\nu') \in \mathfrak{h}^* \times \Lambda$,

$$\widehat{\pi}_{\lambda,\nu} \cong \widehat{\pi}_{\lambda',\nu'} \iff \exists w \in W : (\lambda',\nu') = (w * \lambda, w\nu).$$

7.3. Concluding remarks. We end with a couple of (incomplete) calculations regarding the above analysis, involving central characters.

1. It is natural to ask if the central character associated to an irreducible admissible module V, determines its minimal type. Thus, given that $\chi_V = \chi(\mu_1, \mu_2)$, how does one determine the minimal type of V?

   It is clear that if $V \cong \widehat{\pi}_{\lambda,\nu}$ (from above results), then

   $$\mu_1 = w_1 * \lambda, \qquad \mu_2 = w_2 * (\nu - \lambda - 2\rho) = w_2\nu - w_2 * \lambda - 2\rho,$$

   for some $w_1, w_2 \in W$. Now note that

   $$\mu_1 + w_1 w_2^{-1} * \mu_2 = w_1 * \lambda + w_1\nu - w_1 * \lambda - 2\rho = w_1\nu - 2\rho.$$

   Hence $\nu_w := 2\rho + \mu_1 + w * \mu_2 \in \Lambda$ for some $w \in W$; moreover, for every such w, $\overline{\nu_w}$ is a candidate for the minimal type, by these calculations. Thus, if w is not uniquely identified from above, then neither is $\overline{\nu}$.

2. Similarly, given $(\lambda,\nu), (\lambda',\nu') \in \mathfrak{h}^* \times \Lambda$, a necessary condition for $\widehat{\pi}_{\lambda,\nu}$ to be isomorphic to $\widehat{\pi}_{\lambda',\nu'}$ is that their central characters and minimal types coincide. It is natural to ask if this data is also sufficient to determine the isomorphism type.

   Clearly, in order to have the same minimal type, Corollary 4.3 implies that $\nu' \in W\nu$. Say $\nu' = w_1\nu$. Now since the infinitesimal characters coincide, Theorem 4.9 implies:

   $$\lambda' = w_2 * \lambda, \qquad \nu' - \lambda' - 2\rho = w * (\nu - \lambda - 2\rho) = w\nu - w * \lambda - 2\rho.$$

   Using these equations translates to the following condition:

   $$w_1\nu - w_2 * \lambda = w\nu - w * \lambda,$$

and this data may not have the unique solution: $w_1 = w_2 = w$.

The reason for this discrepancy is Equation (2.8): the representation $\widehat{\pi}_{\lambda,\nu}$ carries the same data as its minimal type and the action of Ω on it. The above data only accounts for the minimal type and the action of the proper subset $Z(\mathfrak{U}(\mathfrak{g} \times \mathfrak{g})) \subsetneq \Omega$. For instance, $Z(\mathfrak{U}\overline{\mathfrak{g}})$ is not accounted for.

To conclude, we have tried to explain the flavour of some of the results in [PRV2], as well as their connection to, and impact on, subsequent research in a wide variety of directions in the field. From the multiplicity problem and obtaining components in tensor products of finite-dimensional modules, to PRV determinants and annihilators of Verma modules, to the classification of all irreducible admissible modules as in Harish-Chandra's grand program on semisimple Lie groups - the work [PRV2] has contributed to, and inspired much subsequent research in, many aspects of representation theory. The list of results and connections mentioned in this article is by no means complete, but we hope that it suffices to convince the reader of the importance and influence of this work in representation theory.

Acknowledgments

I greatly thank Professor Shrawan Kumar for his generous help in pointing out several references and follow-up results in the literature. I would also like to thank very much Professor Dipendra Prasad for his time and patience in answering several of my questions, and Professor Vyjayanthi Chari for useful references and discussions. An excellent historical account of the writing of [PRV2] can be found in Professor V.S. Varadarajan's reminiscences [Va], and this was of great help in the writing of the present work as well.

References

[BB]   A.A. Beilinson and J.N. Bernstein, *Localisation de* $\mathfrak{g}$-*modules*, C. R. Acad. Sci. Paris, Ser. 1 **292** (1981), 15–18.

[BZ]   A. Berenstein and A. Zelevinsky, *Tensor product multiplicities, canonical bases and totally positive varieties*, Inventiones Mathematicae **143** (2001), 77–128.

[BGG]   J. Bernstein, I.M. Gelfand, and S.I. Gelfand, *A category of* $\mathfrak{g}$ *modules*, Functional Analysis and Applications **10** (1976), 87–92.

[Br]   R. Brauer, *Sur la multiplication des caractéristiques des groups continus et semi-simples*, C. R. Acad. Sci. Paris **204** (1937), 1784–1786.

[Bru]   F. Bruhat, *Sur les réprésentations induites des groupes de Lie*, Bull. Soc. Math. France **84** (1956), 97–205.

[CG1]   V. Chari and J. Greenstein, *Current algebras, highest weight categories and quivers*, Advances in Mathematics **216** (2007), no. 2, 811–840.

[CG2]   ———, *Minimal affinizations as projective objects*, Journal of Geometry and Physics **61** (2011), no. 3, 594–609.

[CKR]   V. Chari, A. Khare, and T.B. Ridenour, *Faces of polytopes and Koszul algebras*, Journal of Pure and Applied Algebra **216** (2012), no. 7, 1611–1625.

[CP]   V. Chari and A. Pressley, *A new family of irreducible,integrable modules for affine Lie algebras*, Mathematische Annalen **277** (1987), no. 3, 543–562.

[DR]   I. Dimitrov and M. Roth, *Geometric realization of PRV components and the Littlewood-Richardson cone*, Contemporary Mathematics **490**: Symmetry in Mathematics and Physics, D. Babbitt, V. Chari, and R. Fioresi, Eds. (2009), 83–95.

[Du1]   M. Duflo, *Construction of primitive ideals in an enveloping algebra*, Lie groups and their representations (1971 János Bolyai Math. Soc. Summer School in Mathematics, Budapest), I.M. Gelfand, Ed. (1975), 77–93.

[Du2]   ———, *Représentations irréductibles des groupes semi-simples complexes*, Springer Lecture Notes in Mathematics **497** (1975), 26–88.

[Du3]   ———, *Sur la classification des idéaux primitifs dans l'algèbre enveloppante d'une algèbre de Lie semi-semisimple*, Ann. Math. **105** (1977), 107–120.

[FL]   D.R. Farkas and G. Letzter, *Quantized representation theory following Joseph*, Progress in Mathematics **243**, Part I: Studies in Lie Theory (2006), 9–17.

[GN]   I.M. Gelfand and M.A. Naimark, *Unitary representations of the classical groups*, Trudy Mat. Inst. Steklov **36**, Moscow-Leningrad, 1950.

[GL]   M. Gorelik and E. Lanzmann, *The annihilation theorem for Lie superalgebra* $\mathfrak{osp}(1, 2\ell)$, Inventiones Mathematicae **137** (1999), 651–680.

[Hal]   B.C. Hall, *Lie groups, Lie algebras, and representations: an elementary introduction*, Graduate Texts in Mathematics, no. **222**, Springer-Verlag, Berlin-New York, 2004.

[Har1]   Harish-Chandra, *On some applications of the universal enveloping algebra of a semi-simple Lie algebra*, Trans. Amer. Math. Soc. **70** (1951), 28–96.

[Har2]   ———, *Representations of a semi-simple Lie group on a Banach space: I*, Trans. Amer. Math. Soc. **75** (1953), 185–243.

[Har3]   ———, *Representations of semi-simple Lie groups: II*, Trans. Amer. Math. Soc. **76** (1954), 26–65.

[Har4]   ———, *The Plancherel formula for complex semi-simple Lie groups*, Trans. Amer. Math. Soc. **76** (1954), 485–528.

[Hay]   M. Hayashi, *The moduli space of SU(3)-flat connections and the fusion rules*, Proc. Amer. Math. Soc. **127** (1999), 1545-1555.

[Hu1]   J.E. Humphreys, *Introduction to Lie algebras and representation theory*, Graduate Texts in Mathematics, no. **9**, Springer-Verlag, Berlin-New York, 1972.

[Hu2]   _____, *Representations of semisimple Lie algebras in the BGG Category* $\mathcal{O}$, Graduate Studies in Mathematics **94**, American Mathematical Society, Providence, RI, 2008.

[Jo1]   A. Joseph, *Quantum groups and their primitive ideals*, Ergeb. Math. Grenzgeb. (3) **29**, Springer, Berlin, 1995.

[Jo2]   _____, *A completion of the quantized enveloping algebra of a Kac-Moody algebra*, Journal of Algebra **214** (1999), no. 1, 235–275.

[Jo3]   _____, *On the Kostant-Parthasarathy-Ranga Rao-Varadarajan determinants, I. Injectivity and multiplicities*, Journal of Algebra **241** (2001), 27–45.

[JL1]   A. Joseph and G. Letzter, *Separation of variables for quantized enveloping algebras*, American Journal of Mathematics **116** (1994), 127–177.

[JL2]   _____, *Verma modules annihilators for quantized enveloping algebras*, Ann. Ecole Norm. Sup. **28** (1995), 493–526.

[JL3]   _____, *On the Kostant-Parthasarathy-Ranga Rao-Varadarajan determinants, II. Construction of the KPRV determinants*, Journal of Algebra **241** (2001), 46–66.

[JLT]   A. Joseph, G. Letzter, and D. Todoric, *On the Kostant-Parthasarathy-Ranga Rao-Varadarajan determinants, III. Computation of the KPRV determinants*, Journal of Algebra **241** (2001), 67–88.

[JT]   A. Joseph and D. Todoric, *On the quantum KPRV determinants for semisimple and affine Lie algebras*, Algebras and Representation Theory **5** (2002), 57–99.

[Ka]   M. Kashiwara, *On crystal bases*, Canadian Math. Soc. Conf. Proc. **16** (1995), 155-197.

[KLV]   D. Kazhdan, M. Larsen, and Y. Varshavsky, *The Tannakian Formalism and the Langlands Conjectures*, preprint, http://arxiv.org/abs/1006.3864.

[Kh]   A. Khare, *Axiomatic framework for the BGG Category* $\mathcal{O}$, preprint, http://arxiv.org/abs/0811.2080.

[Kl]   A.U. Klimyk, *On the multiplicities of weights of representations and the multiplicities of representations of semisimple Lie algebras*, Dokl. Acad. Nauk SSSR **177** (1967), 1001–1004.

[KZ]   A.W. Knapp and G. Zuckerman, *Classification of irreducible tempered representations of semisimple groups*, Ann. Math. **116** (1982), 389–455.

[Ko1]   B. Kostant, *A formula for the multiplicity of a weight*, Trans. Amer. Math. Soc. **93** (1959), 53–73.

[Ko2]   _____, *Lie group representations on polynomial rings*, American Journal of Mathematics **85** (1963), no. 3, 327–404.

[Ko3]   _____, *On the existence and irreducibility of certain series of representations*, Lie groups and their representations (1971 János Bolyai Math. Soc. Summer School in Mathematics, Budapest), I.M. Gelfand, Ed. (1975), 231–331.

[Ko4]   _____, *Clifford algebra analogue of the Hopf-Koszul-Samelson Theorem, the ρ-decomposition $C(\mathfrak{g}) = \operatorname{End} V_\rho \otimes C(P)$, and the $\mathfrak{g}$-module structure of $\wedge \mathfrak{g}$*, Advances in Mathematics **125** (1997), 275–350.

[Ku1]   S. Kumar, *Proof of the Parthasarathy-Ranga Rao-Varadarajan conjecture*, Inventiones Mathematicae **93** (1988), 117–130.

[Ku2]   _____, *Existence of certain components in the tensor product of two integrable highest weight modules for Kac-Moody algebras*, Advanced series in Mathematical Physics **7**: Infinite dimensional Lie algebras and groups, V.G. Kac, Ed. (1989), 25–38.

[Ku3]   _____, *A refinement of the PRV conjecture*, Inventiones Mathematicae **97** (1989), 305–311.

[Ku4]   _____, *Proof of Wahl's conjecture on surjectivity of the Gaussian map for flag varieties*, American Journal of Mathematics **114** (1992), 1201–1220.

[Ku5]   _____, *Tensor Product Decomposition*, Proceedings of the International Congress of Mathematicians (2010).

[La]   R.P. Langlands, *On the classification of irreducible representations of real algebraic groups*, Mathematical Surveys and Monographs **31**: Representation theory and harmonic analysis on semisimple Lie groups, P.J. Sally and D.A. Vogan, Eds. (1989), 101–170.

[Le]   J. Lepowsky, *Algebraic results on representations of semisimple Lie groups*, Trans. Amer. Math. Soc. **176** (1973), 1–44.

[LMC]   J. Lepowsky and G.W. McCollum, *On the determination of irreducible modules by restriction to a subalgebra*, Trans. Amer. Math. Soc. **176** (1973), 45–57.

[Li]   P. Littelmann, *A Littlewood-Richardson rule for symmetrizable Kac-Moody algebras*, Inventiones Mathematicae **116** (1994), 329–346.

[Lu]   G. Lusztig, *Canonical bases arising from quantized enveloping algebras II*, Prog. Theor. Phys. **102** (1990), 175–201.

[Ma1]   O. Mathieu, *Construction d'un groupe de Kac-Moody et applications*, Compositio Mathematica **69** (1989), no. 1, 37–60.

[Ma2]   _____, *Classification of Harish-Chandra modules over the Virasoro Lie algebra*, Inventiones Mathematicae **107** (1992), 225–234.

[Ma3]   _____, *Classification of irreducible weight modules*, Annales de l'institut Fourier **50** (2000), no. 2, 537–592.

[MPR1] P.L. Montagard, B. Pasquier, and N. Ressayre, *Two generalizations of the PRV conjecture*, Compositio Mathematica **147** (2011), no. 4, 1321–1336.

[MPR2] ———, *Generalizations of the PRV conjecture, II*, preprint, http://arxiv.org/abs/1110.4621.

[MP] R.V. Moody and A. Pianzola, *Lie algebras with triangular decompositions*, Canadian Mathematical Society Series of Monographs and Advanced Texts, Wiley Interscience, New York-Toronto, 1995.

[PY] D.I. Panyushev and O.S. Yakimova, *The PRV-formula for tensor product decompositions and its applications*, Functional Analysis and Applications **42** (2008), no. 1, 45–52.

[PRV1] K.R. Parthasarathy, R. Ranga Rao, and V.S. Varadarajan, *Representations of complex semisimple Lie groups and Lie algebras*, Bull. Amer. Math. Soc. **72** (1966), 522–525.

[PRV2] ———, *Representations of complex semisimple Lie groups and Lie algebras*, Ann. Math. **85** (1967), 383–429.

[Po] P. Polo, *Variétés de Schubert et excellentes filtrations*, Astérisque (Orbites unipotentes et représentations) **173-174** (1989), 281–311.

[Ra] K.N. Rajeswari, *Standard monomial theoretic proof of PRV conjecture*, Communications in Algebra **19** (1991), 347–425.

[Sh] N.N. Shapovalov, *On a bilinear form on the universal enveloping algebra of a complex semisimple Lie algebra*, Functional Analysis and Its Applications **6** (1972), 307–312.

[St] R. Steinberg, *A general Clebsch-Gordan Theorem*, Bull. Amer. Math. Soc. **67** (1961), 406–407.

[Va] V.S. Varadarajan, *Some mathematical reminiscences*, Methods and Applications of Analysis **9** (2002), no. 3, v–xviii.

[Ve] D.N. Verma, *Structure of certain induced representations of complex semisimple Lie algebras*, Bull. Amer. Math. Soc. **74** (1968), no. 1, 160–166.

[Vo] D.A. Vogan Jr., *The algebraic structure of the representation of semisimple Lie groups. I*, Ann. Math. **109** (1979), no. 1, 1–60.

[YZ] C.A.S. Young and R. Zegers, *Dorey's rule and the q-characters of simply-laced quantum affine algebras*, Commun. Math. Phys. **302** (2011), 789–813.

[Zh1] D.P. Zhelobenko, *The analysis of irreducibility in the class of elementary representations of a complex semisimple Lie group*, Math. USSR Izv. **2** (1968), no. 1, 105–128.

[Zh2] ———, *Harmonic analysis on complex semisimple Lie groups* (Russian), Mir, Moscow, 1974.

Parthasarathy Dirac Operators and Discrete Series Representations

S. Mehdi*

I first met with Professor Rajagopalan Parthasarathy in September 2003 in Mumbai at Tata Institute of Fundamental Research, where I was about to spend, along with my family, a semester as a research assistant. With great patience and availability, Professor R. Parthasarathy, and his wife, made my stay both pleasant and fruitful. I have learned a lot from him about representation theoretic Dirac operators and I feel quite lucky I have learned this topic directly from its source. Professor Rajagopalan Parthasarathy has always been a thoughtful guide and an inexhaustible source of inspiration to me.

1. Introduction

Symmetries have always been a natural source of inspiration and a strong guiding principle for both scientists and artists, as it is remarkably discussed in a book by Hermann Weyl [53]. In particular, the notions of group of symmetries and group actions lie at the heart of deep concepts both in physics and in mathematics. In this context, Dirac operators provide elegant and powerful bridges between quantum physics, geometry and representation theory. In particular, using representation theoretic Dirac operators, Parthasarathy provided deep and elegant answers to each of the following important representation theoretic questions.

(1) Find *explicit constructions* of representations of a Lie group G.

(2) Find a *criteria for unitarizability* of representations of G.

(3) *Classify* representations of G.

(4) Describe *topological invariants* of locally symmetric spaces in terms of representations of G.

---

*Université de Lorraine, Institut Elie Cartan de Lorraine, UMR 7502, Metz, F-57045, France CNRS, Institut Elie Cartan de Lorraine, UMR 7502, Metz, F-57045, France. E-mail: mehdi@univ-metz.fr

In the present paper, we shall focus on problem (1) in the context of discrete series representations and discuss some aspects of the impact of Parthasarathy's Dirac operators on representation theory of real reductive Lie groups. We mention that an expanded version which includes a thorough discussion on Parthasarathy's contributions to the four problems stated above is in preparation.

Let $\mathcal{H}$ be a Hilbert space, i.e a complex vector space equipped with a Hermitian inner product $\langle\,,\rangle$ which is complete with respect to the associated norm $\|\,v\,\| = \langle v,v\rangle^{1/2}$. $U(\mathcal{H})$ denotes the group of unitary operators on $\mathcal{H}$, i.e invertible linear operators $T : \mathcal{H} \to \mathcal{H}$ satisfying $\langle T(v), T(w)\rangle = \langle v,w\rangle$ for all $v, w \in \mathcal{H}$, and G is a (real) Lie group, i.e a topological group equipped with a structure of smooth (real) manifold for which the product and the inverse are smooth maps. A unitary representation of G is a group homomorphism $\pi : G \to U(\mathcal{H})$ such that the map $g \mapsto \langle\pi(g)v, w\rangle$ is continuous for every $v, w \in \mathcal{H}$. A unitary representation is irreducible if it has no proper invariant closed subspace, and a matrix coefficient of the representation $(\pi, \mathcal{H})$ is a map $g \mapsto \langle\pi(g)v, w\rangle$, for some fixed non zero vectors v and w in $\mathcal{H}$. An important example of unitary representation is the regular representation, i.e the action by left (or right) translations on the Hilbert space $L^2(G)$ of square integrable complex functions on G with respect to Haar measure. Two unitary representations $(\pi_1, \mathcal{H}_1)$ and $(\pi_2, \mathcal{H}_2)$ of G are (unitarily) equivalent if there exists a (unitary) invertible linear operator $T : \mathcal{H}_1 \to \mathcal{H}_2$ such that $\pi_2(g) \circ T = T \circ \pi_1(g)$ for all $g \in G$. The unitary dual of G is the set $\widehat{G}$ of equivalence classes of irreducible unitary representations of G. In particular, the unitary dual of an abelian Lie group reduces to unitary characters, i.e one dimensional unitary representations. On the other hand, every irreducible unitary representation of a compact Lie group is finite dimensional, and every finite dimensional representation of a compact Lie group is equivalent to a unitary representation. Moreover, every matrix coefficient of an irreducible unitary representation $(\pi, \mathcal{H})$ of a compact Lie group G lies in $L^2(G)$. Write $L^2(G)_\pi$ for the $\dim(\pi)$-dimensional G-invariant subspace of $L^2(G)$ spanned by the matrix coefficients of π (here $\dim(\pi) = \dim(\mathcal{H})$ is the degree of π). In other words, the restriction of the regular representation of G to $L^2(G)_\pi$ can be decomposed as the direct sum of $\dim(\pi)$ irreducible representations, each of which is equivalent to π. By Peter-Weyl theorem, $L^2(G)$ decomposes into the direct sum $L^2(G) = \bigoplus_{\pi \in \widehat{G}} L^2(G)_\pi$ of irreducible representations, and each $\pi \in \widehat{G}$ occurs with multiplicity equals to $\dim(\pi)$ (see [21]).

For non compact Lie groups the situation is rather different and much more complicated. Indeed such a group may possess infinite dimensional irreducible representations which could be unitary or not. Moreover, the only irreducible finite dimensional unitary representation of a non compact simple Lie group is

131

the trivial representation. This is the case, for example, for the special linear group $SL(2, \mathbf{R})$ of 2×2 real matrices with determinant one. Unlike compact groups, representation theory of non compact semisimple Lie groups is not all well understood. However, the class of discrete series representations is well understood and plays a fundamental role in harmonic analysis. A *discrete series representation* of a connected semisimple real Lie group (with finite center) G is an irreducible unitary representation π on a Hilbert space $\mathcal{H}$ which occurs in the regular representation of G on $L^2(G)$, i.e one of its matrix coefficients lies in $L^2(G)$. In this case, every matrix coefficient of π is square integrable. Discrete series representations are exactly the representations that occur discretely in the decomposition of the regular representation. In particular, when G is compact all irreducible unitary representations representations of G belong to the discrete series. The discrete series representations may be thought of as generalization of the compact situation. Discrete series representations do not always exist, and Harish-Chandra proved that G has discrete series representations if, and only if, G admits a compact Cartan subgroup (see [12], [13], [14]). When the discrete series is empty, i.e there are no irreducible representations occurring discretely as summands in $L^2(G)$, the Plancherel formula is purely an integral. This is the case of complex groups (see [11]). Recall that the Plancherel formula for G amounts to an explicit decomposition of the regular representation of G into a direct integral of irreducible unitary representations of G (see [21]). As we shall see in the next sections, Harish-Chandra has completely classified the set of equivalences classes of discrete series representations using characters of maximal tori. The remaining important problem about discrete series representations, raised by Langlands in [27], was to provide their explicit realization, i.e find a homogeneous bundle $\mathcal{F} \to G/H$ over a homogeneous space G/H and a G-invariant differential operator acting on sections of this bundle such that the space on which the representation acts coincides with the kernel of this differential operator. This was the first of several important contributions of Parthasarathy to representation theory.

More precisely, the introduction by Atiyah and Singer of Dirac operators in the context of Riemannian geometry and index theory (see [3]), extended by Bott to the case of homogeneous spaces (see [6]), naturally led to the introduction of Dirac operators in representation theory. In his Ph.D. thesis, under the supervision of M. S. Narasimhan at TIFR (Mumbai), Parthasarathy used a (geometric) G-invariant Dirac operator

$$D_{G/K}(\mathcal{E}) : L^2(G/K, \mathcal{S} \otimes \mathcal{E}) \to L^2(G/K, \mathcal{S} \otimes \mathcal{E})$$

acting on L^2-sections of some twisted bundle $\mathcal{S} \otimes \mathcal{E}$ over G/K of the spin bundle, where K is a maximal compact subgroup of maximal rank in a real semisimple Lie group G, and proved that the kernel of $D_{G/K}(\mathcal{E})$ is a discrete series representation of G (see [39]). This provided a beautiful answer to problem (1)

132

above and was the first step towards a complete construction by Atiyah and Schmid (see [2]) of discrete series representations of real semisimple Lie groups (besides the original monumental and indispensable work of Harish-Chandra on discrete series). It turned out that Parthasarathy's construction provides a strong guiding principle leading to the construction of various series of representations for a larger class of groups (see Section 5). Parthasarathy's construction of discrete series representations may be viewed as a non compact analog of the Borel-Weil-Bott theorem (see Section 2) in the case when the complex structure on G/K is replaced by a (twisted) spin structure, and the Dirac operator replaces the sum of the Dolbeault operator plus its formal adjoint. We mention that Harish-Chandra obtained in the 1950's (infinite dimensional) versions of the Borel-Weil-Bott theorem (see [47]). This will be the topic of Section 4. For the sake of completeness, Section 3 contains a survey of the main ingredients of Harish-Chandra's classification of discrete series representations, needed to describe Parthasarathy's construction. Moreover, as a motivation, we have included basic material on Borel-Weil-Bott theorem and Dolbeault cohomology in Section 2. However, we assume that the reader is familiar both with the theory of semisimple Lie algebras and their finite dimensional representations (see, e.g., [20]) and with basic concepts of representation theory of semisimple Lie groups (see, e.g., [21]).

It turns out that the introduction by Parthasarathy of representation theoretic Dirac operators had a tremendous impact on the development of representation theory of real reductive Lie groups and was the source of various rich applications and tools such as, just to mention a few: *Parthasarathy-Dirac inequality, unitarizability and classification of irreducible* $(\mathfrak{g}, K)$*-modules, Dirac cohomology, cohomologically induced modules, vanishing theorems for Betti numbers of locally symmetric spaces.* This will be briefly discussed in the last section. The diagram on next page pictures our discussion and describes the emergence of Parthasarathy Dirac operators.

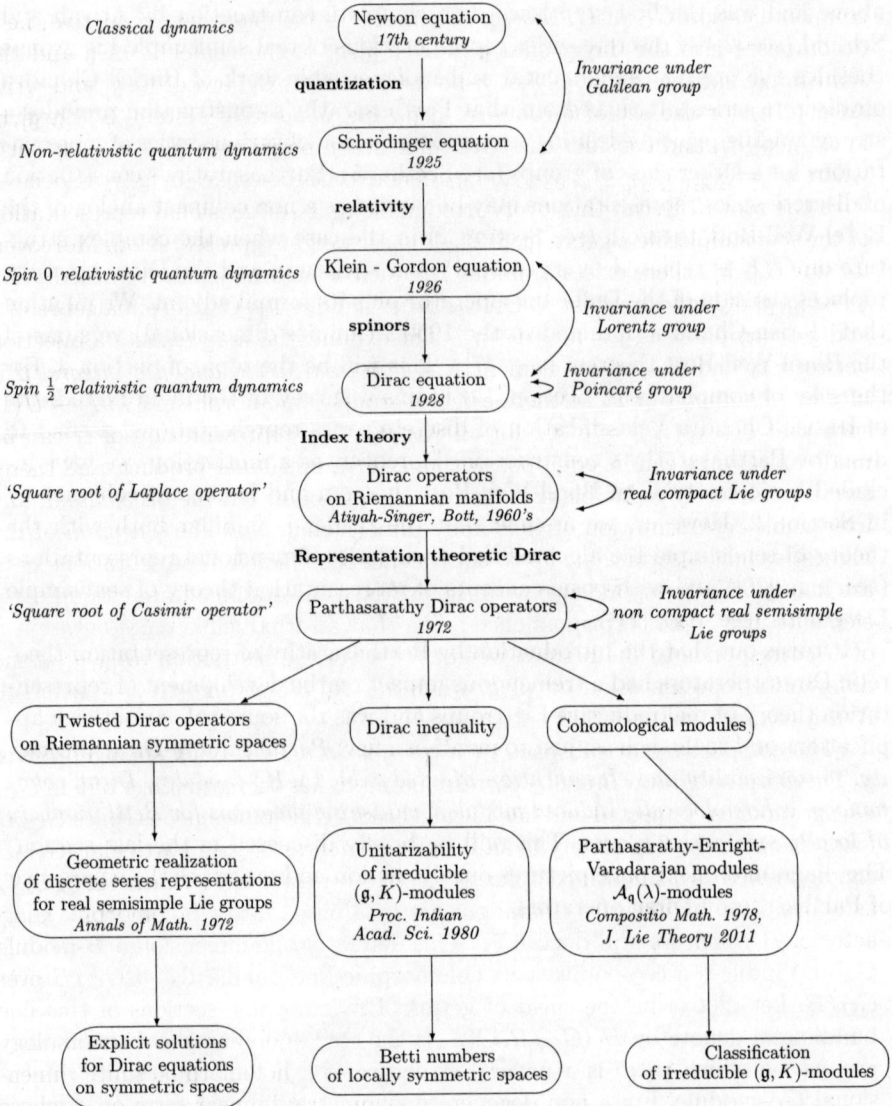

2. The Borel-Weil-Bott theorem

Let G be a compact connected real Lie group and T a maximal torus in G, i.e T is a compact connected abelian Lie subgroup of G, with Lie algebras $\mathfrak{g}_0$ and $\mathfrak{t}_0$ respectively. The complexification $\mathfrak{g}$ of $\mathfrak{g}_0$ is the Lie algebra of the complexification $G_{\mathbf{C}}$ of G. Moreover, since G is compact, $\mathfrak{g}$ is reductive, i.e $\mathfrak{g}$ decomposes into the direct sum $\mathfrak{g} = \mathcal{Z} + [\mathfrak{g}, \mathfrak{g}]$ of the center $\mathcal{Z}$ of $\mathfrak{g}$ and the semisimple subalgebra $[\mathfrak{g}, \mathfrak{g}]$. Let $\mathfrak{h}$ be a Cartan subalgebra of $[\mathfrak{g}, \mathfrak{g}]$ and write $\Delta([\mathfrak{g}, \mathfrak{g}], \mathfrak{h})$ for the set of $\mathfrak{h}$-roots in $[\mathfrak{g}, \mathfrak{g}]$. Fix a positive system $\Delta^+([\mathfrak{g}, \mathfrak{g}], \mathfrak{h})$ in $\Delta([\mathfrak{g}, \mathfrak{g}], \mathfrak{h})$, and consider the triangular decomposition $[\mathfrak{g}, \mathfrak{g}] = \mathfrak{n}^- + \mathfrak{h} + \mathfrak{n}^+$, where $\mathfrak{n}^{\pm} = \sum_{\alpha \in \Delta^+([\mathfrak{g},\mathfrak{g}],\mathfrak{h})} \mathfrak{g}_{\pm\alpha}$ and $\mathfrak{g}_\beta$ denotes the root space in $[\mathfrak{g}, \mathfrak{g}]$ associated with the root β. Recall that every irreducible finite dimensional representation of the semisimple Lie algebra $[\mathfrak{g}, \mathfrak{g}]$ is a highest weight representation. Moreover, for any $\lambda \in \mathfrak{h}^\star$, there exists a unique irreducible highest weight representation V_λ of $[\mathfrak{g}, \mathfrak{g}]$ of highest weight λ. On the other hand, $\mathfrak{t}$ is a Cartan subalgebra of $\mathfrak{g}$ and $\mathfrak{g}$ decomposes as $\mathfrak{g} = \mathfrak{n}^- + \mathfrak{t} + \mathfrak{n}^+$. The exponential map realizes a bijection between $\mathfrak{t}_0$ and T, and an element $\lambda \in \mathfrak{t}^\star$ is said to be *analytic integral* if it lifts to a character of T, i.e there exists a group homomorphism $\xi_\lambda : T \to U(1)$ such that $\xi_\lambda(\exp(X)) = e^{\lambda(X)}$ for all $X \in \mathfrak{t}_0$. Any representation of G can be made unitary, i.e can be equipped with a G-invariant inner product, and therefore is completely reducible if it is finite dimensional. On the other hand, any irreducible representation of G is finite dimensional. Now the Theorem of the highest weight asserts that, up to equivalence, irreducible representations of G are in one-one correspondence with the $\Delta^+([\mathfrak{g}, \mathfrak{g}], \mathfrak{h})$-dominant analytic integral elements in $\mathfrak{t}^\star$. The correspondence being that an irreducible representation π of G corresponds to the highest weight $\lambda \in \mathfrak{t}^\star$ of the restriction to $[\mathfrak{g}, \mathfrak{g}]$ of the differential representation $d\pi$ of $\mathfrak{g}$, where $d\pi(X) = \frac{d}{dt}\pi(\exp(tX))\big|_{t=0}$ for all $X \in \mathfrak{g}$ (see, for instance, [20], [21]).

The complex vector subspace $\mathfrak{t} + \mathfrak{n}^+$ of $\mathfrak{g}$ is actually a Borel subalgebra $\mathfrak{b}$ of $\mathfrak{g}$. The Borel subgroup B of $G_{\mathbf{C}}$ corresponding to $\mathfrak{b}$ is the normalizer of $\mathfrak{b}$ in $G_{\mathbf{C}}$. The group B is connected and has Lie algebra $\mathfrak{b}$. The homogeneous space $G_{\mathbf{C}}/B$ identifies with the flag variety of all Borel subalgebras of $\mathfrak{g}$. In particular, $G_{\mathbf{C}}/B$ can be viewed as a smooth complex projective variety. Now a character ξ_λ of T, as above, lifts, by linearity (making it act by 1 on $\mathfrak{n}^+$), to a holomorphic character of B which we still denote ξ_λ. This defines a one-dimensional B-module $\mathbf{C}_\lambda$ and induces a $G_{\mathbf{C}}$-equivariant holomorphic line bundle $\mathcal{L}_\lambda \to G_{\mathbf{C}}/B$ over $G_{\mathbf{C}}/B$. Let $\mathcal{O}(\mathcal{L}_\lambda)$ be the sheaf of germs of holomorphic sections of this line bundle and denote by $H^{\cdot}(G_{\mathbf{C}}/B, \mathcal{O}(\mathcal{L}_\lambda))$ the corresponding sheaf cohomology groups, on which there is a natural structure of a holomorphic finite dimensional $G_{\mathbf{C}}$-module. Fix a non degenerate symmetric bilinear form on $\mathfrak{g}$ whose restriction to $[\mathfrak{g}, \mathfrak{g}]$ coincides with the Killing form and consider the weight lattice $\{\lambda \in \mathfrak{h}^\star \mid \langle \lambda, \check{\alpha} \rangle \in \mathbf{Z}$ for all $\alpha \in \Delta([\mathfrak{g}, \mathfrak{g}], \mathfrak{h})\}$ of $[\mathfrak{g}, \mathfrak{g}]$, where $\check{\alpha} = 2\dfrac{\alpha}{\langle \alpha, \alpha \rangle}$. Writing ρ for half the sum of the members in $\Delta^+([\mathfrak{g}, \mathfrak{g}], \mathfrak{h})$ and using the diffeomorphism $G/T \sim G_{\mathbf{C}}/B$, the Borel-Weil-Bott theorem may be stated as follows.

Theorem 2.1. *(Borel-Weil-Bott Theorem, see [46])*
For $\lambda \in R$, one has:

$$H^k(G/T, \mathcal{O}(\mathcal{L}_\lambda)) = \begin{cases} 0 \text{ if } \lambda + \rho \text{ is singular or if } k \neq \sharp\{\alpha \in \Delta^+([\mathfrak{g}, \mathfrak{g}], \mathfrak{h}) \mid \langle \lambda + \rho, \alpha \rangle < 0\}. \\ \\ V_{w(\lambda+\rho)-\rho} \text{ otherwise, where } w \text{ is the unique element in } W \text{ which} \\ \quad makes w(\lambda + \rho) - \rho \text{ dominant with respect to } \Delta^+([\mathfrak{g}, \mathfrak{g}], \mathfrak{h}). \end{cases}$$

When λ is dominant, i.e $w = e$, one recovers the *Borel-Weil theorem*. More precisely, if $\Gamma_{hol}(G/T, \mathcal{L}_\lambda)$ denotes the finite dimensional space G-module of holomorphic sections of the line bundle $\mathcal{L}_\lambda \to G/T$ over G/T, then $H^0(G/T, \mathcal{O}(\mathcal{L}_\lambda)) \simeq \Gamma_{hol}(G/T, \mathcal{L}_\lambda)$.

Theorem 2.2. *(Borel-Weil Theorem, see [46])*
For $\lambda \in R$, one has:

$$\Gamma_{hol}(G/T, \mathcal{L}_\lambda) = \begin{cases} V_\lambda \text{ if } \lambda \text{ is dominant}, \\ \\ 0 \text{ otherwise}. \end{cases} \tag{2.3}$$

There is an interesting relation between Borel-Weil-Bott theorem and *Dolbeault cohomology*. Let $A^k(G/T, \mathcal{L}_\lambda)$ be the space of smooth $\mathcal{L}_\lambda$-valued forms of type $(0, k)$, i.e the space of smooth sections of $\mathcal{L}_\lambda \otimes \Lambda^k(T^{(0,1)}G/T)^\star$. The Dolbeault operator $\overline{\partial} : A^k(G/T, \mathcal{L}_\lambda) \to A^{k+1}(G/T, \mathcal{L}_\lambda)$ satisfies $\overline{\partial}^2 = 0$, and defines a complex which induces the Dolbeault cohomology spaces

$$H^k_{\overline{\partial}}(G/T, \mathcal{L}_\lambda) = \ker(\overline{\partial})/\mathrm{im}(\overline{\partial}) \tag{2.4}$$

which are finite dimensional G-modules, since G is compact and $\overline{\partial}$ is G-invariant. The Dolbeault theorem asserts that Dolbeault cohomology $H^\cdot_{\overline{\partial}}(G/T, \mathcal{L}_\lambda)$ is isomorphic, as a G-module, to the sheaf cohomology $H^\cdot(G/T, \mathcal{O}(\mathcal{L}_\lambda))$.

Theorem 2.5. *(Dolbeault Theorem, see [46])*

$$H^k_{\overline{\partial}}(G/T, \mathcal{L}_\lambda) \simeq H^k(G/T, \mathcal{O}(\mathcal{L}_\lambda)). \tag{2.6}$$

Let $\overline{\partial}^\star$ be the formal adjoint of $\overline{\partial}$, and form the G-invariant Laplace-Beltrami operator on $A^k(G/T, \mathcal{L}_\lambda)$

$$\square \overset{def.}{=} \overline{\partial}\overline{\partial}^\star + \overline{\partial}^\star\overline{\partial} = (\overline{\partial} + \overline{\partial}^\star)^2 : A^k(G/T, \mathcal{L}_\lambda) \to A^k(G/T, \mathcal{L}_\lambda) \tag{2.7}$$

so that one has the isomorphism of G-modules:

$$H^k(G/T, \mathcal{O}(\mathcal{L}_\lambda)) \simeq \ker(\square). \tag{2.8}$$

In other words, combining Borel-Weil-Bott and Dolbeault theorems, one deduces that all irreducible representations of G can be realized as some Dolbeault cohomology space (2.4), or equivalently as the kernel of some Laplace-Beltrami operator (2.8).

3. Harish-Chandra Classification of the Discrete Series

Let G be a connected real semisimple Lie group with finite center and K a maximal compact subgroup K of G, write $\mathfrak{g}_0$ and $\mathfrak{k}_0$ for the Lie algebras of G and K respectively. One of the fundamental insights of Harish-Chandra is to reduce the study of irreducible unitary representations of G to the study of irreducible $(\mathfrak{g}, K)$-modules. This is a combination of representation theory of compact Lie groups, which is well understood, with the algebraic theory of modules for the enveloping algebra $U(\mathfrak{g})$ of the complexification $\mathfrak{g}$ of $\mathfrak{g}_0$. More precisely, following [48], recall that a $(\mathfrak{g}, K)$-module is a complex vector space V together with representations $\mathfrak{g} \to \text{End}(V)$ and $K \to GL(V)$ denoted by $\cdot$ such that:

(i) V is the (algebraic) direct sum of finite-dimensional irreducible representations of K,

(ii) the infinitesimal action of K on G agrees with the restriction of $\mathfrak{g} \to \text{End}(V)$,

(iii) for $X \in \mathfrak{g}$ and $k \in K$, we have $k \cdot (X \cdot (k^{-1} \cdot v)) = (\text{Ad}(k)(X)) \cdot v$.

A $(\mathfrak{g}, K)$-module V is said to be *finitely generated* if it is finitely generated as a $U(\mathfrak{g})$-module, and it is said to be *irreducible* if the only $\mathfrak{g}$ and K-invariant subspaces of V are V and $\{0\}$. If V and W are $(\mathfrak{g}, K)$-modules then we denote by $\text{Hom}_{\mathfrak{g},K}(V, W)$ the space of all $\mathfrak{g}$-homomorphisms that are also K-homomorphism of V to W, then V and W are said to be *equivalent* if there is an invertible element in $\text{Hom}_{\mathfrak{g},K}(V, W)$. A $(\mathfrak{g}, K)$-module X is unitary if X is equipped with a positive-definite scalar product:

$$\langle k \cdot v, k \cdot w \rangle = \langle v, w \rangle \text{ and } \langle X \cdot v, w \rangle = -\langle v, X \cdot w \rangle.$$

Let π be a representation of G on a Hilbert space $\mathcal{H}$. A vector $v \in \mathcal{H}$ is called *K-finite* if the space $\text{span}_{k \in K}\{\pi(k)v\}$ is finite dimensional. A vector $v \in \mathcal{H}$ is called a *smooth vector* if the map $g \mapsto \pi(g)v$ is a smooth map. The space of smooth vectors is a dense G-invariant subspace of $\mathcal{H}$. One nice feature of smooth vectors is that they induce, by differentiation, a representation $d\pi$

of $U(\mathfrak{g})$, which is some sort of *linearization* of the representation π. The representation π is admissible if $\pi(K)$ operates by unitary operators and if each τ in $\widehat{K}$ occurs with only finite multiplicities in the restriction $\pi_{|_K}$ of π to K. Harish-Chandra proved that irreducible unitary representations are admissible. Two admissible representations are said to be infinitesimally equivalent if their associated $(\mathfrak{g}, K)$-modules are isomorphic. One interesting feature of admissible representations is that their K-finite vectors are automatically smooth (the converse is not true), and the space of K-finite vectors is a $\mathfrak{g}$-invariant dense subspace. It is known that an admissible representation π of G on $\mathcal{H}$ is irreducible if, and only if, its associated $U(\mathfrak{g})$-module $\mathcal{H}^K$ is irreducible. The $U(\mathfrak{g})$-module $\mathcal{H}^K$ is a $(\mathfrak{g}, K)$-module, known as the *Harish-Chandra module* of π. If X is an irreducible unitary $(\mathfrak{g}, K)$-module, then X is the Harish-Chandra module of a unique irreducible unitary representation of G. Therefore, to classify discrete series representations, Harish-Chandra suggested to classify the corresponding Harish-Chandra modules by their infinitesimal characters.

Let π an irreducible admissible representation of G on a Hilbert space $\mathcal{H}$. If $Z(\mathfrak{g})$ denotes the center of $U(\mathfrak{g})$, then $d\pi(X)$ is a scalar multiple of the identity map $\mathrm{Id}_{\mathcal{H}^K}$ for every $X \in Z(\mathfrak{g})$. We obtain a homomorphism

$$\chi_\pi : Z(\mathfrak{g}) \to \mathbf{C}$$

such that $d\pi(X) = \chi_\pi(X)\mathrm{Id}_{\mathcal{H}^K}$ for all $X \in Z(\mathfrak{g})$, known as the *infinitesimal character* of π. Any irreducible admissible representation of G admits an infinitesimal character, so does a discrete series representation of G. The idea of Harish-Chandra is to identify $Z(\mathfrak{g})$ with an algebra whose characters are easier to manage. Fix a Cartan subalgebra $\mathfrak{t}$ of $\mathfrak{g}$, by the Poincaré-Birkhoff-Witt theorem, one has

$$\mathcal{U}(\mathfrak{g}) = \mathcal{U}(\mathfrak{t}) \oplus \left(\mathfrak{n}^-\mathcal{U}(\mathfrak{g}) + \mathcal{U}(\mathfrak{g})\mathfrak{n}^+\right)$$

and we let

$$\widetilde{\zeta} : \mathcal{Z}(\mathfrak{g}) \to \mathcal{U}(\mathfrak{t})$$

be the projection on the first factor. Viewing $\mathcal{U}(\mathfrak{t})$ as the algebra of polynomial functions on the vector dual $\mathfrak{t}^\star$ of $\mathfrak{t}$, one defines the ρ-*shift* map

$$T_{\rho(\mathfrak{g})} : \mathcal{U}(\mathfrak{t}) \to \mathcal{U}(\mathfrak{t})$$

by $(T_{\rho(\mathfrak{g})}f)(\lambda) = f(\lambda - \rho(\mathfrak{g})) \quad \forall f \in \mathcal{U}(\mathfrak{t}), \ \forall \lambda \in \mathfrak{t}^\star$, and the *Harish-Chandra map*

$$\zeta : \mathcal{Z}(\mathfrak{g}) \to \mathcal{U}(\mathfrak{t})$$

given by $\zeta = T_{\rho(\mathfrak{g})} \circ \widetilde{\zeta}$. An important feature of the Harish-Chandra map is that, if W is the Weyl group associated with $\mathfrak{h}$-roots in $\mathfrak{g}$ then ζ is an algebra isomorphism from $\mathcal{Z}(\mathfrak{g})$ onto the algebra $\mathcal{U}(\mathfrak{t})^W$ of polynomial functions on $\mathfrak{t}^\star$ which are W-invariant:

$$\mathcal{Z}(\mathfrak{g}) \simeq S(\mathfrak{t})^W.$$

For $\lambda \in \mathfrak{t}^\star$, write

$$\zeta_\lambda : \mathcal{Z}(\mathfrak{g}) \to \mathbf{C}$$

for the composition map of ζ with the evaluation at λ. Two homomorphisms ζ_λ and $\zeta_{\lambda'}$ coincide if, and only if, λ and λ' are conjugate under W. An infinitesimal character, like every homomorphism from $\mathcal{Z}(\mathfrak{g})$ to $\mathbf{C}$, is of the form ζ_λ for some $\lambda \in \mathfrak{t}^\star$. In particular, the characters of $Z(\mathfrak{g})$ are in 1-1 correspondence with a half-open Weyl chamber in $\mathfrak{t}^\star$.

We can now state *Harish-Chandra classification of discrete series representations*. First, let G be a connected real semisimple Lie group with finite center and K a maximal compact subgroup of G. Write $\mathfrak{g}$ (resp. $\mathfrak{k}$) for the complexification of the Lie algebra of G (resp. K). Suppose there exists a Cartan subalgebra $\mathfrak{t}$ of $\mathfrak{g}$ which is contained in $\mathfrak{k}$. Let $\lambda \in \mathfrak{t}^\star$ be non singular relative to $\Delta(\mathfrak{g}, \mathfrak{t})$ (i.e $\langle \lambda, \alpha \rangle \neq 0$ for all $\alpha \in \Delta(\mathfrak{g}, \mathfrak{t})$) and fix the positive system $\Delta^+(\mathfrak{g}, \mathfrak{t})$ defined as $\{\alpha \in \Delta(\mathfrak{g}, \mathfrak{t}) \mid \langle \lambda, \alpha \rangle > 0\}$. Then $\Delta^+(\mathfrak{k}, \mathfrak{t}) \overset{def.}{=} \Delta^+(\mathfrak{g}, \mathfrak{t}) \cap \Delta(\mathfrak{k}, \mathfrak{t})$ is a positive system for $\mathfrak{t}$-roots in $\mathfrak{k}$. Denote by $W_\mathfrak{g}$ (resp. $W_\mathfrak{k}$) the Weyl group of $\Delta(\mathfrak{g}, \mathfrak{t})$ (resp. $\Delta(\mathfrak{k}, \mathfrak{t})$) and $\rho(\mathfrak{g})$ (resp. $\rho(\mathfrak{k})$) half the sum of the elements in $\Delta^+(\mathfrak{g}, \mathfrak{t})$ (resp. $\Delta^+(\mathfrak{k}, \mathfrak{t})$).

Theorem 3.1. *(Harish-Chandra, see [12], [13], [14])*
The set of (equivalence classes of) discrete series representations of G is parametrized by the set

$$\mathcal{HC}_{ds}(G) = \left\{ \lambda \in \mathfrak{t}^\star \mid \langle \lambda, \check{\alpha} \rangle \in \mathbf{Z} \text{ and } \langle \lambda, \check{\alpha} \rangle \neq 0 \, \forall \alpha \in \Delta(\mathfrak{g}, \mathfrak{t}), \text{ and } \langle \lambda, \check{\alpha} \rangle > 0 \, \forall \alpha \in \Delta^+(\mathfrak{k}, \mathfrak{t}) \right\}.$$

For $\lambda \in \mathcal{HC}_{ds}(G)$, the discrete series representation π_λ of G is characterized by the following properties:

(1) π_λ has infinitesimal character ζ_λ,

(2) $\pi_\lambda \mid_K$ contains with multiplicity one the K-type with highest weight $\Lambda = \lambda + \rho(\mathfrak{g}) - 2\rho(\mathfrak{k})$,

(3) if Λ' is the highest weight of a K-type in $\pi_\lambda \mid_K$, then $\Lambda' = \Lambda + \sum_{\alpha \in \Delta^+(\mathfrak{g}, \mathfrak{t})} n_\alpha \alpha$ for integers $n_\alpha \geq 0$.
Furthermore two such representations π_λ are equivalent if, and only if, their parameters λ are conjugate under the Weyl group of K. In particular the number of discrete series representations of G with a fixed infinitesimal character is the (finite) number of elements in $W_\mathfrak{g}/W_\mathfrak{k}$.

λ is called the *Harish-Chandra parameter* of the discrete series representation π_λ and Λ is its *Blattner parameter* (see [21]). Discrete series representations are also characterized by their Harish-Chandra character which is the analog of the usual character of finite dimensional representations of a compact group. Namely, suppose that π is an admissible representation of G on a Hilbert space $\mathcal{H}$. If f is a compactly supported function on the group G, then the operator on $\mathcal{H}$ defined by $\pi(f) \overset{def.}{=} \int_G f(g)\pi(g)dg$ is of trace class, and the distribution $\Theta_\pi : f \mapsto \mathrm{Tr}(\pi(f))$ is called the *character* (or *Harish-Chandra character*) of π. Θ_π is a distribution on G which is invariant under conjugation and is an eigendistribution of the center of the universal enveloping algebra of G, with eigenvalue the infinitesimal character of the representation π. Harish-Chandra proved that Θ_π, actually any invariant eigendistribution on G, is given by a locally integrable function and π is determined up to infinitesimal equivalence by its character Θ_π. Letting T be a compact Cartan subgroup of G with complexified Lie algebra $\mathfrak{t}$, Harish-Chandra computed the restriction to the set $T^{reg} = \{t \in T \mid \alpha(t) \neq 1 \ \forall \alpha \in \Delta(\mathfrak{g}, \mathfrak{t})\}$ of regular elements in T of characters of discrete series representations.

Theorem 3.2. *(Harish-Chandra, see [12], [13], [14])*

$$\Theta_{\pi_\lambda}(t) = (-1)^{\frac{1}{2}\dim(G/K)} \frac{\sum_{w \in W_{\mathfrak{k}}} \epsilon(w) t^{w(\lambda)}}{\Pi_{\alpha \in \Delta^+(\mathfrak{k}, \mathfrak{t})}(t^{\alpha/2} - t^{-\alpha/2})} \ for \ \lambda \in \mathcal{H}C_{ds}(G), \ t \in T^{reg}.$$

Note that when G is compact, π_λ is the irreducible highest representation of G with highest weight $\lambda - \rho(\mathfrak{g})$ and infinitesimal character ζ_λ. In this case, Θ_{π_λ} reduces to the usual Weyl character formula for the irreducible finite dimensional representation of G with highest weight $\lambda - \rho(\mathfrak{g})$.

4. Parthasarathy Dirac Operators and Construction of Discrete Series Representations

Besides Harish-Chandra's classification, it is also useful to have an explicit realization of discrete series representations. More precisely, let G be a connected real semisimple Lie group and H be a closed subgroup of G. By an explicit geometric realization of a representation π of G in a Hilbert space $\mathcal{H}$ we mean finding a homogeneous bundle $\mathcal{F} \to G/H$ over G/H and a G-invariant differential operator acting on sections of this bundle such that $\mathcal{H}$ equals (or rather lies in) the kernel of this differential operator. For example, when G is compact all irreducible representations of G belong to the discrete series of G, and the Borel-Weil-Bott theorem (2.3) realizes the discrete series as spaces of holomorphic sections of line bundles over G/T, where T is a maximal torus in G. Furthermore, the Dolbeault theorem realizes the discrete series as Dolbeault cohomology space (2.4), or equivalently as the kernel of some Laplace-Beltrami operator (2.8). When G is non compact, K is a maximal compact subgroup of G and G/K is Hermitian, i.e the Riemannian symmetric space of the non compact type G/K is equipped with a complex structure for which G acts with holomorphically, Narasimhan and Okamoto proved an analog of the Borel-Weil-Bott theorem which realizes most of the discrete series representations of G (see [36]). In [23] and [26] a generalization of the Borel-Weil-Bott Theorem is also proved. See also [47]. Let T be a Cartan subgroup of G which is a maximal torus of K and let $\mathfrak{s}$ be the complexification of the orthogonal complement $\mathfrak{s}_0$ of $\mathfrak{k}_0$ in $\mathfrak{g}_0$ with respect to the Killing form. There is a $\mathrm{Ad}(K)$-invariant almost complex structure J on $\mathfrak{s}_0$ such that $\mathfrak{s}$ splits into $\mathfrak{s} = \mathfrak{s}^+ + \mathfrak{s}^-$ with $J_{|_{\mathfrak{s}^\pm}} = \pm\sqrt{-1}\,\mathrm{Id}$. One may view $\mathfrak{s}^+$ (resp. $\mathfrak{s}^-$) as the subspace of holomorphic (resp. antiholomorphic) tangent vectors at eK in G/K. Fix a positive system P for t-roots in $\mathfrak{g}$ such that $P_n \subset P$ where $\mathfrak{s}^\pm = \sum_{\alpha \in P_n} \mathfrak{g}_{\pm\alpha}$. Then $P = P_k \cup P_n$ where P_k is a positive system for t-roots in $\mathfrak{k}$. Let $\mathcal{C}$ be the subset of the weight lattice of $\mathfrak{g}$ given by

$$\mathcal{C} = \{\lambda \in \mathfrak{t}^\star \mid \langle \lambda, \check{\alpha}\rangle \in \mathbf{Z} \text{ and } \langle \lambda + \rho(\mathfrak{g}), \alpha\rangle \neq 0 \ \forall \alpha \in \Delta(\mathfrak{g},\mathfrak{t}),$$
$$\text{and } \langle \lambda + \alpha, \alpha\rangle > 0 \ \forall \alpha \in \Delta^+(\mathfrak{k},\mathfrak{t})\}.$$

As in section 2, write V_λ for the highest weight representation of K with highest weight $\lambda \in \mathcal{C}$ with respect to P_k and $V_\lambda \to G/K$ for the corresponding holomorphic homogeneous (unitary) vector bundle induced on G/K. Write $\square_\lambda$ for the Laplace-Beltrami operator (2.7) acting on the completion of the space of square integrable V_λ-valued $(0,p)$-forms on G/K and $\ker(\square_\lambda)$ for the kernel of $\square_\lambda$. Narasimhan and Okamoto proved that for $\lambda \in \mathcal{C}$ 'sufficiently regular' the G-module $\ker(\square_\lambda)$ is a discrete series representation of G with Harish-Chandra parameter λ (see [36]). Here 'λ sufficiently regular' means that there exists a positive constant C such that $\langle \lambda + \rho(\mathfrak{g}), \alpha\rangle > C$ for all $\alpha \in P_n$. More generally, they also obtained a series of unitary representations such that every discrete series representation of G is a direct summand of an element of this series. The concrete construction of discrete series by Narasimhan and Okamoto uses in a crucial way the existence of a complex structure on G/K. However, for a general semisimple Lie group G, a complex structure on G/K need not exist. In this context, and in analogy to Narasimhan and Okamoto, Hotta has constructed 'most' of the discrete series of a non compact semisimple Lie group G with finite center as certain eigenspaces of Casimir operators rather than Dolbeault cohomology spaces (see [16]). We also mention Schmid's realization of discrete series representations in terms of L^2-cohomology (see [45], [46]).

In order to treat all non-compact real semisimple Lie groups rather than just the ones whose symmetric space is Hermitian, Narasimhan and Kostant suggested the idea to replace Dolbeault operator plus its formal adjoint by some twisted Dirac operator. In particular, Parthasarathy realized discrete series rep-

141

resentations as spaces of square integrable vector-valued harmonic spinors on the homogeneous space G/K. More precisely, let G be a non compact connected real semisimple Lie group with finite center and K a maximal compact subgroup of G of maximal rank. Let $\mathfrak{s}_0$ be the $\mathrm{Ad}(K)$-invariant orthogonal complement of the Lie algebra $\mathfrak{k}_0$ in $\mathfrak{g}_0$ with respect to the Killing form $\langle,\rangle$. The restriction to $\mathfrak{s}_0$ of the Killing form remains non degenerate and extends linearly to a non degenerate K-invariant $\langle,\rangle_{\mathfrak{s}}$ on the complexification $\mathfrak{s}$ of $\mathfrak{s}_0$. The Lie group $SO(\mathfrak{s}_0)$ of orthogonal endomorphisms of $\mathfrak{s}_0$ relative to $\langle\ ,\ \rangle_{\mathfrak{s}}$ has Lie algebra

$$\mathfrak{so}(\mathfrak{s}_0) = \{A \in \mathrm{End}(\mathfrak{s}_0) \mid \langle AX, Y\rangle_{\mathfrak{s}} + \langle X, AY\rangle_{\mathfrak{s}} = 0, \quad \text{for all } X, Y \in \mathfrak{s}_0\}.$$

Observe that, for all X and Y in $\mathfrak{s}_0$, the endomorphisms $R_{X,Y}$ defined by

$$R_{X,Y}(W) = \langle Y, W\rangle_{\mathfrak{s}} X - \langle X, W\rangle_{\mathfrak{s}} Y, \quad \text{for } W \in \mathfrak{s}_0,$$

span $\mathfrak{so}(\mathfrak{s}_0)$. There is an embedding of $\mathfrak{so}(\mathfrak{s}_0)$ in the Clifford algebra of $\mathfrak{s}$. Indeed, recall that the Clifford algebra $C(\mathfrak{s})$ of $\mathfrak{s}$ is defined as the quotient of the tensor algebra $T(\mathfrak{s})$ of $\mathfrak{s}$ by the ideal $\mathcal{I}$ generated by elements $X \otimes Y + Y \otimes X - 2\langle X, Y\rangle_{\mathfrak{s}}$, with X and Y in $\mathfrak{s}$:

$$C(\mathfrak{s}) = T(\mathfrak{s})/\mathcal{I}.$$

The linear extension of $R_{X,Y} \mapsto \frac{1}{2}(XY - YX)$ is an injective Lie algebra homomorphism $\mathfrak{so}(\mathfrak{s}) \to C_2(\mathfrak{s})$, where $C_2(\mathfrak{s})$ is the Lie algebra defined as the subspace of $C(\mathfrak{s})$ generated by the degree 2 elements. Since $\mathfrak{s}$ is even dimensional, we may choose two maximal dual isotropic subspaces V and $V^\star$ of $\mathfrak{s}$ with respect to $\langle,\rangle_{\mathfrak{s}}$ such that

$$\mathfrak{s} = V \oplus V^\star.$$

Denote by $\wedge V^\star$ the exterior algebra of $V^\star$, equipped with the interior product $\imath$ and exterior multiplication ϵ, and define a map $\gamma : \mathfrak{s} \to \mathrm{End}(\wedge V^\star)$ by

$$\gamma(v + v^\star)(u) = (\imath(v) + \epsilon(v^\star))(u) \text{ for all elements } u \in \wedge^l V^\star.$$

This map extends naturally to a map, still denoted γ, on the Clifford algebra of $\mathfrak{s}$, and is known as the Clifford multiplication. Finally, if $S_\mathfrak{s}^+ \overset{def.}{=} \sum_{l \text{ even}} \wedge^l V^\star$ and $S_\mathfrak{s}^- \overset{def.}{=} \sum_{l \text{ odd}} \wedge^l V^\star$, then the half spin representations $(\sigma_\mathfrak{s}^\pm, S_\mathfrak{s}^\pm)$ of $\mathfrak{so}(\mathfrak{s}_0)$ is

$$\sigma_\mathfrak{s}^\pm(R_{X,Y}) = \frac{1}{2}[\gamma(X), \gamma(Y)] \text{ for all } X, Y \in \mathfrak{s}_0.$$

Now the 'half spin' representations of $\mathfrak{k}_0$ are defined by

$$s_\mathfrak{s}^\pm = \sigma_\mathfrak{s}^\pm \circ \mathrm{ad}.$$

Note that the connected group $SO(\mathfrak{s}_0)$ of orthogonal automorphisms of $\mathfrak{s}_0$ with Lie algebra $\mathfrak{so}(\mathfrak{s}_0)$ need not act on $S_\mathfrak{s}^\pm$, but its double cover $Spin(\mathfrak{s}_0)$ does. Similarly the group K does not act on $S_\mathfrak{s}^\pm$, but its spin double cover $\widetilde{K}$ does (see [18]). So let (τ, E) be a finite dimensional representation of $\mathfrak{k}_0$ such that the $\mathfrak{k}_0$-representations $S_\mathfrak{s}^\pm \otimes E$ integrate to representations $\beta^\pm$ of K. It induces smooth homogeneous vector bundles $\mathcal{S}_\mathfrak{s}^\pm \otimes \mathcal{E}$ over G/K. The space of corresponding smooth sections

$$
\begin{aligned}
C^\infty(G/K, \mathcal{S}_\mathfrak{s}^\pm \otimes \mathcal{E}) \quad &\overset{def.}{=} \quad \{f : G \to S_\mathfrak{s}^\pm \otimes E \mid f \text{ is smooth and} \\
&\qquad f(gk) = \beta^\pm(k^{-1})(f(g)), \text{ for } g \in G, k \in K\} \\
&\overset{G\text{-modules}}{\simeq} \quad \{C^\infty(G) \otimes (S_\mathfrak{s}^\pm \otimes E)\}^K
\end{aligned}
$$

is a G-module under left translations, where K acts as $R \otimes \beta^\pm$ and R denotes the right translation on $C^\infty(G)$. Fixing an orthonormal basis $\{X_j\}$ of $\mathfrak{s}_0$, a K-invariant element in $\mathcal{U}(\mathfrak{g}) \otimes \mathrm{Hom}_\mathbf{C}(S_\mathfrak{s}^\pm \otimes E, S_\mathfrak{s}^\mp \otimes E)$ is defined by $\sum_j X_j \otimes (\gamma(X_j) \otimes 1)$. Letting $\mathcal{U}(\mathfrak{g})$ acts by left invariant differential operator:

$$
(R(X)f)(g) \overset{def.}{=} \frac{d}{dt}f(g\exp(tX))|_{t=0} \text{ for } X \in \mathfrak{g},
$$

Parthasarathy defined the following G-invariant differential operators

$$
D_{G/K}^\pm(\mathcal{E}) : C^\infty(G/K, \mathcal{S}_\mathfrak{s}^\pm \otimes \mathcal{E}) \to C^\infty(G/K, \mathcal{S}_\mathfrak{s}^\mp \otimes \mathcal{E})
$$

by

$$
D_{G/K}^\pm(\mathcal{E}) = \sum_j R(X_j) \otimes \gamma(X_j) \otimes 1 \tag{4.1}
$$

and also provided a formula for its square (see [39]):

$$
D_{G/K}^\mp(\mathcal{E}) \circ D_{G/K}^\pm(\mathcal{E}) = R(\Omega_G) \otimes 1 \otimes 1 + 1 \otimes s_\mathfrak{s}^\pm(\Omega_K) \otimes 1 - 1 \otimes 1 \otimes \tau(\Omega_K) \tag{4.2}
$$

acting on $\{C^\infty(G) \otimes (S_\mathfrak{s}^\pm \otimes E)\}^K \simeq C^\infty(G/K, \mathcal{S}_\mathfrak{s}^\pm \otimes \mathcal{E})$, where Ω_G (resp. Ω_K) denotes the Casimir operator of G (resp. K). In other words, fixing a Cartan subalgebra $\mathfrak{t}$ of $\mathfrak{g}$ contained in $\mathfrak{k}$, choosing compatible positive systems $\Delta^+(\mathfrak{g}, \mathfrak{t})$ and $\Delta^+(\mathfrak{k}, \mathfrak{t})$ for $\mathfrak{t}$-roots in $\mathfrak{g}$ and $\mathfrak{k}$ respectively, and taking $E = E_\mu$ an irreducible highest weight representation of $\mathfrak{k}_0$ with highest weight μ with respect to $\Delta^+(\mathfrak{k}, \mathfrak{t})$, the square of the operator $D_{G/K}(\mathcal{E}_\mu)$ differs from the Casimir operator by a constant depending only on μ, $\rho(\mathfrak{g})$ and $\rho(\mathfrak{k})$. In particular, the operators $D_{G/K}(\mathcal{E}_\mu)$ are *Dirac operators*, which we shall call *Parthasarathy geometric Dirac operators*. It should be noted that the above expression for $D_{G/K}(\mathcal{E})$ is independent of the basis $\{X_j\}$.

Next, consider the subspaces $C_c^\infty(G/K, \mathcal{S}_s^\pm \otimes \mathcal{E}_\mu)$ of compactly supported smooth sections equipped with the canonical inner product, and write $L^2(G/K, \mathcal{S}_s^\pm \otimes \mathcal{E}_\mu)$ for their completions, i.e G-modules consisting of Hilbert spaces of square integrable sections of the bundles $\mathcal{S}_s^\pm \otimes \mathcal{E}_\mu$ over G/K. The self adjoint extension of $D_{G/K}^\pm(\mathcal{E}_\mu)$ to $L^2(G/K, \mathcal{S}_s^\pm \otimes \mathcal{E}_\mu)$ will still be denoted by the same symbol. Since $D_{G/K}^\pm(\mathcal{E}_\mu)$ are G-invariant operators, their L^2-kernels $\ker_{L^2}(D_{G/K}^\pm(\mathcal{E}_\mu))$ defines unitary G-modules $\delta_\lambda^\pm$. Keeping in mind the set $\mathcal{C}$ defined above and the formula (4.2) for the square of the Dirac operator, *Parthasarathy geometric realization* of (most of) discrete series representations of G may be stated as follows.

Theorem 4.3. *(Parthasarathy, see [39])*

(1) Let $\lambda \in \mathcal{C}$ and p_λ be the number of $\alpha \in \Delta^+(\mathfrak{g}, \mathfrak{t}) \setminus \Delta^+(\mathfrak{k}, \mathfrak{t})$ such that $\langle \lambda + \rho(\mathfrak{g}), \alpha \rangle < 0$, then

$$\Theta_{\delta_{\lambda+\rho(\mathfrak{g})-\rho(\mathfrak{k})}^+} - \Theta_{\delta_{\lambda+\rho(\mathfrak{g})-\rho(\mathfrak{k})}^-} = (-1)^{p_\lambda} \Theta_{\pi_{\lambda+\rho(\mathfrak{g})}}$$

where $\pi_{\lambda+\rho(\mathfrak{g})}$ is the discrete series representation of G with infinitesimal character $\lambda + \rho(\mathfrak{g})$ defined in the previous section.

(2) Let $\lambda \in \mathfrak{t}^\star$ be such that $\langle \lambda, \check{\alpha} \rangle \in \mathbf{Z}$ for all $\alpha \in \Delta(\mathfrak{g}, \mathfrak{t})$ and $\langle \lambda, \alpha \rangle \geq 0$ for all $\alpha \in \Delta^+(\mathfrak{g}, \mathfrak{t})$. There exists a unique element $\sigma \in W_{\mathfrak{g}}$ such that $\Delta^+(\mathfrak{k}, \mathfrak{t}) \subset \sigma(\Delta^+(\mathfrak{g}, \mathfrak{t}))$ and $\sigma(\lambda + \rho(\mathfrak{g})) - \rho(\mathfrak{g}) \in \mathcal{C}$. Then if $\langle \lambda, \alpha \rangle \neq 0$ for all $\alpha \in \Delta^+(\mathfrak{g}, \mathfrak{t}) \setminus \Delta^+(\mathfrak{k}, \mathfrak{t})$, one has

$$\ker_{L^2}(D_{G/K}^{-j(\sigma)}(\mathcal{E}_{\sigma(\lambda+\rho(\mathfrak{g}))-\rho(\mathfrak{k})})) = 0 \text{ for } j(\sigma) = + \text{ (resp. } -)$$
$$\text{if } sgn(\sigma) = 1 \text{ (resp. } -1).$$

(3) Furthermore $\ker_{L^2}(D_{G/K}^{j(\sigma)}(\mathcal{E}_{\sigma(\lambda+\rho(\mathfrak{g}))-\rho(\mathfrak{k})}))$ is a discrete series representation with infinitesimal character $\sigma(\lambda + \rho(\mathfrak{g}))$.

Few years later, Atiyah and Schmid used index theory to remove the above regularity condition on the parameter λ to exhaust all discrete series representations as spaces of (twisted) harmonic spinors over G/K (see [2]). In other words, any discrete series of a non compact connected semisimple real Lie group G with finite center is of the form $\ker_{L^2}(D_{G/K}^+(\mathcal{E}))$ for some twisted spin bundle $\mathcal{S}_s \otimes \mathcal{E}$ over G/K. It should be mentioned that the proof of Atiyah and Schmid did not use Harish-Chandra's existence results for discrete series.

5. Impact and Applications of Parthasarathy Dirac Operators

We now briefly discuss applications and aspects of the impact of Parthasarathy Dirac operators on representation theory of real reductive Lie groups.

(1) Parthasarathy's construction of discrete series representations was the stepping-stone to the geometric realization of (smooth vectors of) other series of representations of real reductive Lie groups G as harmonic spinors over homogeneous spaces for G, see, just to mention a few, [4], [5], [7], [10], [16], [33], [34], [35], [54]. In particular, since (geometric) Dirac operators are G-equivariant, their kernels are G-invariant and one can deduce, via intertwining maps, explicit formulas for (non zero) solutions of twisted Dirac equations on homogeneous spaces. When there is an invariant complex structure on G/K, it can be shown that $D_{G/K}(\mathcal{E}_\lambda)$ co-incides, up to a scaling, with the operator $\bar{\partial} + \bar{\partial}^\star$ (see [15]). In this case, Parthasarathy's construction of discrete series may be thought of both as a non compact analog of Borel-Weil-Bott theorem and a generalization of Narasimhan-Okamoto realization of discrete series representations to the case where the Riemannian symmetric space G/K is not equipped with an invariant complex structure. One can consider harmonic spinors as substitute for holomorphic forms. Since the product of holomorphic forms, whenever it makes sense, is holomorphic (see [42]), it is then natural to consider a 'product' for harmonic spinors. Parthasarathy and I described various instances of such a phenomenon in the context of Kostant's cubic Dirac operators for general connected reductive Lie groups G where K is replaced by some connected reductive closed subgroup, not necessarily compact (see [30]). This is connected to the notions of *Jantzen-Zuckerman translation functors* which consist in tensoring a $(\mathfrak{g}, K)$-module with a finite dimensional representation and then projecting to the eigenspace of a particular infinitesimal character. These functors are fundamental tools in representation theory and are related to the notion of *coherent families*, i.e families of $(\mathfrak{g}, K)$-modules parametrized in a *coherent* way (see [48]). Parthasarathy and I proved that one can relate harmonic spinors for an irreducible representation, a finite dimensional irreducible representation and a third representation which is related to the first two via a Zuckerman translation. Moreover, for some modules, there is a geometric counterpart in terms of a 'product map', i.e harmonic spinors with coefficients in a module are expressed as a linear combination of 'products' of harmonic spinors with coefficients in two other modules (see [30], [31]). This is done in the context of Kostant cubic Dirac operators. These

145

results have been extended recently by Pandžić and myself to arbitrary $(\mathfrak{g}, K)$-modules (see [29]).

(2) Parthasarathy computed the square of (an algebraic version of) his Dirac operator $D_{G/K}(\mathcal{E})$ and derived an inequality on infinitesimal characters (see [41]). More precisely, let X be an irreducible unitary $(\mathfrak{g}, K)$-module where $\mathfrak{g}$ denotes the complexified Lie algebra of G. Letting S be the spin module for the spin double cover $\widetilde{K}$ of K, the positive definite invariant Hermitian form on X induces a positive definite invariant form on the tensor product $X \otimes S$ with respect to which the Dirac operator D_X, acting on $X \otimes S$, is self adjoint. In particular, Parthasarathy proves that if X has infinitesimal character Λ and if λ is the highest weight of a simple $\widetilde{K}$-submodule appearing in $X \otimes S$ then

$$\| \Lambda \| \leq \| \lambda + \rho_c \|$$

where ρ_c denotes half the sum of compact positive roots. Parthasarathy proved that this necessary condition is actually sufficient in the case of highest weight modules with regular infinitesimal character when the group G is Hermitian. The above inequality is known as the *Parthasarathy-Dirac inequality* and turned out to be a powerful tool in representation theory to treat the crucial problems of unitarizability and classification, see, for example, [8], [17], [19], [25], [29], [31], [37], [38], [41], [44], [49], [51]. This inequality was also a key ingredient in the proof by Vogan and Zuckerman of the fact that if V is an irreducible unitary $(\mathfrak{g}, K)$-module, of the same infinitesimal character as a finite dimensional representation F, then, writing $F^\star$ for the dual representation of F, $V \otimes F^\star$ has nonzero $(\mathfrak{g}, K)$-cohomology if, and only if, V is a $\mathcal{A}_{\mathfrak{q}}(\lambda)$-module (see [51]). Parthasarathy-Dirac inequality has also deep applications in spectral geometry, it can be used to derive eigenvalue estimates for eigenvalues of the (cubic) Dirac operator (see, for example, [1]).

(3) In the late 1990's, motivated by Parthasarathy Dirac operator and Dirac inequality, Vogan stated a conjecture which relates the infinitesimal character of a $(\mathfrak{g}, K)$-module X and its *Dirac cohomology $H(X)$* (see [50]). More precisely, the Dirac operator D_X acts as a differential on the $\widetilde{K}$-submodule $\ker(D_X^2)$ of $X \otimes S$ defined by the kernel of the square of D_X. The cohomology corresponding to this differential is the *Dirac cohomology $H(X)$* of X. In particular, when X is unitary or finite dimensional, Dirac cohomology reduces to harmonic spinors, i.e $\ker(D_X)$. Vogan's conjecture may be stated as follows: *if X is irreducible and $H(X)$ contains a $\widetilde{K}$-type of highest weight λ, then the infinitesimal character of X is $\lambda + \rho_c$.* This conjecture is now a theorem proved by Huang and Pandžić (see [17]),

146

which was extended by Kostant in the context of the *cubic Dirac operators* (see [22], [24]). Dirac cohomology is related to other cohomologies. For example, if a $(\mathfrak{g}, K)$-module has non zero $(\mathfrak{g}, K)$-cohomology then it has non zero Dirac cohomology (see [17]). On the other hand, when G is Hermitian, Dirac cohomology can be related to Dolbeault cohomology since, in this case, Dirac operator coincides, up to scaling, with the sum of the Dolbeault operator plus its formal adjoint. It is therefore important to classify $(\mathfrak{g}, K)$-modules with non zero Dirac cohomology. This is rather difficult in general. 'Vogan's conjecture' suggests to use Parthasarathy-Dirac inequality to classify irreducible unitary representations of real reductive Lie groups by their Dirac cohomology. In this context, Salamanca-Riba and Vogan have stated a conjecture replacing the above Parthasarathy-Dirac inequality by the following statement (see [44]): *the real part $Re(\Lambda)$ of Λ belongs to the convex hull of the Weyl group orbit of $\lambda + \rho_c$.* This conjecture implies a classification of unitary representations for real reductive Lie groups, in terms of cohomological parabolic induction. Recently, Parthasarathy and I constructed a family of infinite dimensional modules with non zero (cubic) Dirac cohomology, which includes discrete series representations, cohomological modules $\mathcal{A}_\mathfrak{q}(\lambda)$ and Parthasarathy's generalized Enright-Varadarajan modules (see [30], [31], [32]). Parthasarathy's Enright-Varadarajan modules are $(\mathfrak{g}, K)$-modules constructed by Parthasarathy in the late 1970's (see [40], [52]) as a consequence of a generalization of the Enright-Varadarajan algebraic (infinitesimal) construction of discrete series representations (see [9]). It should be mentioned that Parthasarathy's Enright-Varadarajan modules provide another characterization of the cohomological modules $\mathcal{A}_\mathfrak{q}(\lambda)$ described, in the 1980's, by Vogan and Zuckerman, via their classification of modules with non zero $(\mathfrak{g}, K)$-cohomology (see [51]).

(4) As a byproduct of his Dirac inequality, Parthasarathy obtained a vanishing result for the Betti numbers of locally symmetric spaces. The starting point is the celebrated *Matsushima's formula* which relates *Betti numbers* of locally symmetric spaces, dimension of spaces of *automorphic forms* and $(\mathfrak{g}, \mathfrak{k})$-cohomology groups (see [28]). More precisely, let K denote a maximal compact subgroup of a real semisimple Lie group G with finite center, $\mathfrak{g}$ (resp. $\mathfrak{k}$) the complexified Lie algebra of G (resp. K) and $\mathfrak{g} = \mathfrak{k} + \mathfrak{s}$, the corresponding Cartan decomposition. If $(\pi, \mathcal{H})$ is an irreducible unitary representation of G, write $H^{(\ell)}(\mathfrak{g}, \mathfrak{k}, \pi)$ for the ℓth relative $(\mathfrak{g}, \mathfrak{k})$-cohomology group of π. On the other hand, given a cocompact torsion free discrete subgroup Γ of G, the unitary action (by right translations) of G on the Hilbert space $L^2(\Gamma \backslash G)$ decomposes into a countable direct sum of irreducible unitary representations $(\pi, \mathcal{H})$ with finite multiplici-

ties $m_\Gamma(\pi)$. The Matsushima's formula expresses the ℓth Betti number $\beta_\ell(\Gamma\backslash G/K)$ of the compact locally symmetric space $\Gamma\backslash G/K$ as follows:

$$\beta_\ell(\Gamma\backslash G/K) = \sum_{\pi\in\widehat{G},\,\chi_\pi=\chi_0} m_\Gamma(\pi)\dim H^{(\ell)}(\mathfrak{g},\mathfrak{k},\pi),$$

since relative $(\mathfrak{g},\mathfrak{k})$-cohomology of π vanishes if the infinitesimal character χ_π of π does not coincide with the infinitesimal character χ_0 of the one-dimensional trivial G-module. Furthermore, it is known that $H^{(\ell)}(\mathfrak{g},\mathfrak{k},\pi)$ is isomorphic to $\mathrm{Hom}_{\mathfrak{k}}(\Lambda^\ell\mathfrak{s},\mathcal{H})$ when $\chi_\pi = \chi_0$. In particular, to obtain vanishing results on Betti numbers, it is important to know: when is $\mathrm{Hom}_{\mathfrak{k}}(\Lambda^\ell\mathfrak{s},\mathcal{H})$ non zero? Using Dirac operators arguments, Parthasarathy provided an explicit answer in the case of highest weight modules when G/K is an irreducible Hermitian symmetric space (see [41], [42], [43]). These results were extended by Kumaresan in his P. h. D. thesis under the supervision of Parthasarathy (see [25]). In turn Kumaresan's results were used in a crucial way by Vogan and Zuckerman in their classification of modules with non zero $(\mathfrak{g},\mathfrak{k})$-cohomology (see [51]).

References

[1] I. Agricola, *Connections on naturally reductive spaces, their Dirac operator and homogeneous models in string theory*, Comm. Math. Phys. **232** (2003), no. 3, 535–563.

[2] M. Atiyah and W. Schmid, *A geometric construction of the discrete series for semisimple Lie groups*, Inv. Math. **42** (1977), 1–62.

[3] M. Atiyah and I. M. Singer, *The index of elliptic operators on compact manifolds*, Bull. Amer. Math. Soc. **69** (1963), 422–433.

[4] L. Barchini, Szegö mappings, harmonic forms and Dolbeault cohomology, J. Funct. Anal. **118** (1993), 351–406.

[5] L. Barchini, A. W. Knapp and R. Zierau, *Intertwining operators into Dolbeault cohomology representations*, J. Funct. Anal. **107** (1992), 302–341.

[6] R. Bott, *The index theorem for homogeneous differential operators*, Differential and combinatorial topology (A symposium in honor of Marston Morse), 167–186, Princeton Univ. Press, Princeton, N. J, 1965.

[7] R. W. Donley, *Intertwining operators into cohomology representations for semisimple Lie groups*, J. Funct. Anal. **151** (1997), 138–165.

[8] T. J. Enright and R. Parthasarathy, *A proof of a conjecture of Kashiwara and Vergne*, Noncommutative harmonic analysis and Lie groups (Marseille, 1980), pp. 74-90, Lecture Notes in Math., 880, Springer, Berlin-New York, 1981.

[9] T. J. Enright and V. S. Varadarajan, *On an infinitesimal characterization of the discrete series*, Ann. of Math. **102** (1975), 1–15.

[10] T. J. Enright and N. R. Wallach, *Embeddings of unitary highest weight representations and generalized Dirac operators*, Math. Ann. 307 (1997), no. 4, 627–646.

[11] I. M. Gelfand and M.A. Naimark, *An analog of Plancherel's formula for the complex unimodular group*, Doklady Akademii Nauk SSSR **63** (1948), 609–612.

[12] Harish-Chandra, *Discrete series for semisimple Lie groups. I. Construction of invariant eigendistributions*, Acta Math. **113** (1965), 241–318.

[13] Harish-Chandra, *Discrete series for semisimple Lie groups. II. Explicit determination of the characters*, Acta Math. **116** (1966), 1–111.

[14] Harish-Chandra, *Two theorems on semi-simple Lie groups*, Ann. of Math. (2) **83** (1966), 74–128.

[15] N. Hitchin, *Harmonic spinors.* Advances in Math. **14** (1974), 1–55.

[16] R. Hotta, *On a realization of the discrete series for semisimple Lie groups*, J. Math. Soc. Japan **23** (1971), 384–407.

[17] J.-S. Huang and P. Pandžić, *Dirac cohomology, unitary representations and a proof of a conjecture of Vogan*, J. Amer. Math. Soc. **15** (2002), 185–202.

[18] J.-S. Huang and P. Pandžić, *Dirac operators in representation theory*, Mathematics: theory and applications. Series Editor: N. R. Wallach. Birkhäuser 2006.

[19] J.-S. Huang, P. Pandžić and V. Protsak, *Dirac cohomology of Wallach representations*, Pacific J. Math. **250** (2011), no. 1, 163–190.

[20] J. E. Humphreys, *Introduction to Lie algebras and representation theory.* Second printing, revised. Graduate Texts in Mathematics, 9. Springer-Verlag, New York-Berlin, 1978.

[21] A. W. Knapp, *Representation theory of semisimple groups: An overview based on examples*, Princeton University Press, Vol. 36, 1986.

[22] B. Kostant, *A cubic Dirac operator and the emergence of Euler number multiplets of representations for equal rank subgroups*, Duke Math. J. **100** (1999), 447–501.

[23] B. Kostant, *A generalization of the Bott-Borel-Weil theorem and Euler number multiplets of representations*, Conference Moshé Flato 1999 (Dijon). Lett. Math. Phys. **52** (2000), no. 1, 61–78.

[24] B. Kostant, *Dirac-cohomology for the cubic Dirac operator*, Studies in Memory of Issai Schur (Chevaleret/Rehovot, 2000). Progr. Math. **210**, 69–93, Birkhäuser, 2003.

[25] S. Kumaresan, *On the canonical $\mathfrak{k}$-types in the irreducible unitary $\mathfrak{g}$-modules with non-zero relative Lie algebra cohomology*, Inv. Math. **59** (1980), 1–11.

[26] G. D. Landweber, *Harmonic spinors on homogeneous spaces*, Rep. Theory **4**(2000), 466–473.

[27] R. P. Langlands, *Dimensions of spaces of automorphic forms*, Proc. Sympos. Pure Math. **9** (1966), 253–257. (Available at http://www.sunsite.ubc.ca/DigitalMathArchive/Langlands/pdf/dimen-ps.pdf)

[28] Y. Matsushima, *A formula for the Betti numbers of compact locally symmetric Riemannian manifolds*, J. Differential Geometry **1** (1967), 99–109.

[29] S. Mehdi and P. Pandžić, *Dirac cohomology and translation functors*, Under revision.

[30] S. Mehdi and R. Parthasarathy, *A product for harmonic spinors on reductive homogeneous spaces*, J. Lie Theory **18** (2008), 33–44.

[31] S. Mehdi and R. Parthasarathy, *Representation theoretic harmonic spinors for coherent families*, INSA Platinum Jubilee Special Issue, Indian J. Pure Appl. Math. **41** (2010), 133–144.

[32] S. Mehdi and R. Parthasarathy, *Cubic Dirac cohomology for generalized Enright-Varadarajan modules*. J. Lie Theory **21** (2011), 861–884

[33] S. Mehdi and R. Zierau, *Harmonic spinors on semisimple symmetric spaces*, J. Funct. Anal. **198**(2003), 536–557.

[34] S. Mehdi and R. Zierau, *Principal series representations and harmonic spinors*, Adv. Math. **1999**(2006), 1–28.

[35] S. Mehdi and R. Zierau, *Harmonic spinors on reductive homogeneous spaces*, In preparation.

[36] M. S. Narasimhan and K. Okamoto, *An analogue of the Borel-Weil-Bott theorem for Hermitian symmetric pairs of non-compact type*, Ann. of Math. (2) **91** (1970) 486–511.

[37] P. Pandžić, *Dirac operators and unitarizability of Harish-Chandra modules*, Math. Commun. **15** (2010), no. 1, 273–279.

[38] A. Pantano, A. Paul and S. Salamanca-Riba, *The omega-regular unitary dual of the metaplectic group of rank 2*, Council for African American Researchers in the Mathematical Sciences. Vol. V, 1–47, Contemp. Math., 467, Amer. Math. Soc., Providence, RI, 2008.

[39] R. Parthasarathy, *Dirac operator and the discrete series*, Ann. of Math. **96** (1972), 1–30.

[40] R. Parthasarathy, *A generalization of the Enright-Varadarajan modules*, Compositio Math. **36** (1978), no. 1, 53–73. (Available at http://repository.ias.ac.in/view/fellows/Parthasarathy=3ARajagopalan=3A=3A.html)

[41] R. Parthasarathy, *Criteria for the unitarizability of some highest weight modules*, Proc. Indian Acad. Sci. Sect. A Math. Sci. **89** (1980), no. 1, 1–24. (Available at http://repository.ias.ac.in/view/fellows/Parthasarathy=3ARajagopalan=3A=3A.html)

[42] R. Parthasarathy, *Holomorphic forms in $\Gamma \backslash G / K$ and Chern classes*, Topology **21** (1982), no. 2, 157–178.

[43] R. Parthasarathy, *Unitary modules with non vanishing relative Lie algebra cohomology*, Proceedings of the International Congress of Mathematicians, Vol. 1, 2 (Warsaw, 1983), 905–907, PWN, Warsaw, 1984.

[44] S. Salamanca-Riba and D. A. Vogan, *On the classification of unitary representations of reductive Lie groups*, Ann. of Math. (2) **148** (1998), no. 3, 1067–1133.

[45] W. Schmid, L^2-*cohomology and the discrete series*, Ann. of Math. **103** (1976), no. 2, 375–394.

[46] W. Schmid, *Geometric methods in representation theory*, Lecture notes taken by Matvei Libine. London Math. Soc. Lecture Note Ser., 323, Poisson geometry, deformation quantisation and group representations, pp. 273-323, Cambridge Univ. Press, Cambridge, 2005.

[47] V. S. Varadarajan, *Harish-Chandra, his work, and its legacy*, The mathematical legacy of Harish-Chandra (Baltimore, MD, 1998), 1–35, Proc. Sympos. Pure Math. **68**, Amer. Math. Soc., Providence, RI, 2000.

[48] D. A. Vogan, *Representations of real reductive groups*, Progress in Math. 15. Birkhäuser 1981.

[49] D. A. Vogan, *Unitarizability of certain series of representations*, Ann. of Math. **120** (1984), no. 1, 141–187.

[50] D. A. Vogan, *Dirac operators and unitary representations*, 3 talks at MIT Lie groups seminar, Fall 1997.

[51] D. A. Vogan and G. J. Zuckerman, *Unitary representations with non-zero cohomology*, Comp. Math., **53** (1984), 51–90.

[52] N. R. Wallach, *On the Enright-Varadarajan modules: a construction of the discrete series*, Ann. Sci. Ecole Norm. Sup. **9** (1976), 81–101.

[53] H. Weyl, *Symmetry*, Reprint of the 1952 original. Princeton Sc. Library. Princeton Univ. Press, Princeton, NJ, 1989.

[54] J. A. Wolf, *Partially harmonic spinors and representations of reductive Lie groups*, J. Funct. Anal. **15**(1974), 117–154.

The Rôle of Verma in Representation Theory

Henning Haahr Andersen[*]

1. Introduction

The title of this paper is chosen to echo the title of the paper [44] by Daya-Nand Verma. Our aim is to explain the main results and conjectures there and to describe the impact they have had in representation theory in the years since then.

No doubt, for most mathematicians the first association they have, when the name Verma is mentioned, is to the modules that bear his name. Verma studied these modules in his thesis (Yale 1966, see the announcement in [43]). We will start out with a section where we discuss Verma modules and try to illustrate why they are so important in representation theory. At the same time, this section will fix notation and introduce the key objects that we will be dealing with. As will become evident, the theory presented in the remainder of this article is, to a large extent, modeled after the theory of highest weight modules for complex semisimple Lie algebras.

One of the key ideas that Verma presented in [44] was that an affine Weyl group should play a similar role in modular representation theory to the one played by the ordinary Weyl group in representation theory of complex Lie algebras. Moreover, his "metamathematical conjecture" in the last section of [44] says that the modular representation theory, when properly formulated, is independent of p, the underlying characteristic. We shall explain these two ideas in some detail and point to some later developments in which they played a decisive role. Our presentation is heavily influenced by our own background and interests and with all deliberate hindsight we freely take advantage of many newer ideas and concepts. Along the way we take the opportunity to point to some questions and open problems related to Verma's work.

Clearly, we cannot in this short article give all details. Even worse, we can cover only a few aspects of the immense theory that has emerged since Verma

---

[*]Center for Quantum Geometry of Moduli Spaces, Aarhus University, Building 530, Ny Munkegade, 8000 Aarhus C, DENMARK. E-mail: mathha@qgm.au.dk

presented his ideas. We do not pretend to give fair and proper references to the work by many researchers who have contributed to the theory. Still we hope that our presentation together with the references we provide (the reader is advised to track also the further references given there) will give the readers some broader insight into Verma's ideas and into the ways they have been responsible for shaping the developments in representation theory over the last forty years.

2. Verma Modules

2.1. Notation and background.

Let $\mathfrak{g}$ denote a semisimple complex Lie algebra. If $\mathfrak{b}$ is a Borel subalgebra of $\mathfrak{g}$ and $\lambda : \mathfrak{b} \to \mathbb{C}$ is a linear character of $\mathfrak{b}$ then we define

$$\Delta(\lambda) = U(\mathfrak{g}) \otimes_{U(\mathfrak{b})} \mathbb{C}_\lambda. \tag{2.1}$$

Here $U(\mathfrak{g})$ and $U(\mathfrak{b})$ denote the universal enveloping algebras of $\mathfrak{g}$ and $\mathfrak{b}$, respectively. Moreover, $\mathbb{C}_\lambda$ denotes the 1-dimensional $\mathfrak{b}$-module defined by λ. The $\mathfrak{g}$-structure on $\Delta(\lambda)$ comes from left multiplication on the first factor.

The module $\Delta(\lambda)$ in (2.1) was dubbed by Dixmier [19] the *Verma module* with highest weight λ. Verma was not the first to consider these constructions. For instance, Harish-Chandra [23] had studied such modules several years before Verma in his thesis, cf. [43], explored their deeper properties by investigating homomorphisms between them.

Actually, he used an alternative description of $\Delta(\lambda)$, namely as the quotient of $U(\mathfrak{g})$ by the left ideal generated by $\mathfrak{n}$ together with $\{h - \lambda(h) \mid h \in \mathfrak{h}\}$. Here, $\mathfrak{n}$ and $\mathfrak{h}$ are respectively, the nilpotent part and the Cartan subalgebra of $\mathfrak{b}$ (so that $\mathfrak{b} = \mathfrak{h} \oplus \mathfrak{n}$). Moreover, we have used the "modern" notation $\Delta(\lambda)$ for the Verma module of highest weight λ. In the literature, this module is also often denoted $M(\lambda)$, see e.g. [30], [26]. Our notation hints at the fact that a Verma module has unique simple quotient. In more general situations (like general highest weight categories, cf. [17]), modules with similar properties play a corresponding role and are called *standard modules*.

Let R denote the root system for $(\mathfrak{g}, \mathfrak{h})$ and choose the set of positive roots R_+ to be the roots of $\mathfrak{n}$. Write W for the Weyl group of R. It is generated by the reflections s_α, $\alpha \in R_+$. The coroot associated with $\alpha \in R$ is denoted $\alpha^\vee$. The natural action of W on $\mathfrak{h}^*$ is then given by $s_\alpha \lambda = \lambda - \langle \lambda, \alpha^\vee \rangle \alpha$, $\alpha \in R_+, \lambda \in \mathfrak{h}^*$. In addition, we shall consider the ρ-shifted action where ρ denotes half the sum of the positive roots

$$w \cdot \lambda = w(\lambda + \rho) - \rho, \quad \text{for all } \lambda \in \mathfrak{h}^* \tag{2.2}$$

This has become known as the *"dot − action"*.

153

2.2. Verma's theorem.

Theorem 2.1. *[43] If $\alpha \in R_+$ and $\lambda \in \mathfrak{h}^*$ satisfy $\langle \lambda + \rho, \alpha^\vee \rangle \in \mathbb{N}$ then, up to scalars, there is a unique non-trivial $\mathfrak{g}$-homomorphism $\Delta(s_\alpha \cdot \lambda) \to \Delta(\lambda)$.*

For two arbitrary weights $\lambda, \mu \in \mathfrak{h}^*$ any non-zero homomorphism $\Delta(\lambda) \to \Delta(\mu)$ is injective (because $U(\mathfrak{g})$ is a domain). The fact, that there is an inclusion $\Delta(s_\alpha \cdot \lambda) \subset \Delta(\lambda)$ in the case where α is a simple root satisfying the assumption in the theorem, is an easy $\mathfrak{sl}_2$-computation. In that case the weight space $\Delta(\lambda)_{s_\alpha \cdot \lambda}$ is 1-dimensional so the uniqueness is also easy. However, to establish the existence and uniqueness for a general positive root is much more delicate, see [43] or compare [26].

2.3. Linkage.

Let $Z \subset U(\mathfrak{g})$ denote the center of $U(\mathfrak{g})$. For any $\mathfrak{g}$-module M multiplication by an element from Z is then a $\mathfrak{g}$-endomorphism of M. When M is a Verma module it must therefore be a scalar. Hence we get for each $\lambda \in \mathfrak{h}^*$ a character $\chi_\lambda : \mathfrak{h} \to \mathbb{C}$ determined by

$$zm = \chi_\lambda(z)m, \ m \in \Delta(\lambda), \ z \in Z. \tag{2.3}$$

Harish-Chandra proved [23]

Theorem 2.2. *Let $\lambda, \mu \in \mathfrak{h}^*$. Then $\chi_\lambda = \chi_\mu$ if and only if $\lambda \in W \cdot \mu$.*

We shall call this result *the linkage principle*. It implies clearly that $\mathrm{Hom}_\mathfrak{g}(\Delta(\lambda), \Delta(\mu)) = 0$ unless $\lambda \in W \cdot \mu$. Verma was the first to establish the existence and uniqueness of such homomorphisms in non-trivial cases. He went on to formulate a conjecture (discussed below) giving the precise condition for the existence of homomorphisms between two Verma modules.

2.4. Strong linkage.

Let us write $\lambda \uparrow \mu$ if there exists $\alpha \in R_+$ such that $\langle \mu + \rho, \alpha^\vee \rangle \in \mathbb{N}$ and $\lambda = s_\alpha \cdot \mu$.

Definition 2.3. Let $\lambda, \mu \in \mathfrak{h}^*$. We say that λ is *strongly linked* to μ if there exists a chain $\lambda = \lambda_1 \uparrow \lambda_2 \uparrow \cdots \uparrow \lambda_r = \mu$.

Note that if λ is strongly linked to μ we clearly have $\lambda \in W \cdot \mu$. **Verma's conjecture** can now be stated as follows

Conjecture 2.4. ([43], Conjecture 1) Let $\lambda, \mu \in \mathfrak{h}^*$. Then $\Delta(\lambda) \subset \Delta(\mu)$ if and only if λ is strongly linked to μ.

Of course Verma's Theorem (Theorem 2.1 above) proves the "if part" of this conjecture. The "only if part" was established a few years later by Bernstein, Gelfand and Gelfand [8]. We shall call this result *the strong linkage principle*.

2.5. Category $\mathcal{O}$ and its Grothendieck group. In [8] and [9] the authors introduced the category $\mathcal{O}$. This became a much studied and important category in a lot of later work, see e.g. the recent book by Humphreys [26]. In this section we recall the definition of $\mathcal{O}$ and give some of its basic properties.

If M is a $\mathfrak{g}$-module and $\lambda \in \mathfrak{h}^*$ then we set

$$M_\lambda = \{m \in M \mid hm = \lambda(h)m \text{ for all } h \in \mathfrak{h}\}.$$

We call this the λ-*weight space* of M, and if this space is non-zero we say that λ is a *weight* of M. We call M a *weight module* if $M = \bigoplus_{\lambda \in \mathfrak{h}^*} M_\lambda$.

The category $\mathcal{O}$ is then the subcategory of the category of all finitely generated $U(\mathfrak{g})$-modules M for which M is a weight module and M is locally $U(\mathfrak{n})$-finite. The last condition means that for any vector $m \in M$ the subspace $U(\mathfrak{n})m \subset M$ is finite dimensional.

It is not hard to see that all Verma modules from Section 2.1 belong to $\mathcal{O}$. So do all submodules and quotients of Verma modules. Now each Verma module $\Delta(\lambda)$ has a unique simple quotient which we shall call $L(\lambda)$. In fact, the family $\{L(\lambda)\}_{\lambda \in \mathfrak{h}^*}$ is up to isomorphisms the set of simple modules in $\mathcal{O}$.

Among the basic properties of modules in $\mathcal{O}$ is the fact that they have finite composition series [8]. Let $K(\mathcal{O})$ denote the Grothendieck group of $\mathcal{O}$ and write $[M]$ for the image in $K(\mathcal{O})$ of a module $M \in \mathcal{O}$. Then the family $([L(\lambda)])_{\lambda \in \mathfrak{h}^*}$ constitutes a $\mathbb{Z}$-basis of $K(\mathcal{O})$ and we have for any $M \in \mathcal{O}$,

$$[M] = \sum_\lambda [M : L(\lambda)][L(\lambda)]. \tag{2.4}$$

Here $[M : L(\lambda)] \in \mathbb{N}$ is the composition factor multiplicity of $L(\lambda)$ in M. In particular, when M is a Verma module the equation (2.4) reads

$$[\Delta(\lambda)] = \sum_\mu [\Delta(\lambda) : L(\mu)][L(\mu)], \ \lambda \in \mathfrak{h}^*, \tag{2.5}$$

where the sum extends over all $\mu \leq \lambda$ (in the usual ordering $\leq$ on $\mathfrak{h}^*$ given by R_+). We have $[\Delta(\lambda) : L(\lambda)] = 1$ for all λ. In other words, the infinite matrix $[\Delta(\lambda) : L(\mu)])_{\lambda,\mu \in \mathfrak{h}^*}$ is lower triangular (relative to this ordering) with 1's on the diagonal. This means that we may "invert" (2.5) and write

$$[L(\lambda)] = \sum_\mu (L(\lambda) : \Delta(\mu))[\Delta(\mu)], \ \lambda \in \mathfrak{h}^* \tag{2.6}$$

for some $(L(\lambda) : \Delta(\mu)) \in \mathbb{Z}$. Moreover, these coefficients satisfy

$$(L(\lambda) : \Delta(\mu)) = 0 \text{ unless } \mu \leq \lambda \text{ and } (L(\lambda) : \Delta(\lambda)) = 1. \tag{2.7}$$

By the linkage principle (Theorem 2.2) we have, furthermore, $[\Delta(\lambda) : L(\mu)] = 0 = (L(\lambda) : \Delta(\mu))$ unless $\mu \in W \cdot \lambda$ so that the sum in (2.6) is finite. By the strong linkage principle we even get

155

Corollary 2.5. $[\Delta(\lambda) : L(\mu)] = 0 = (L(\lambda) : \Delta(\mu))$ *unless μ is strongly linked to λ.*

2.6. Dual Verma modules.

We have a duality functor $D : \mathcal{O} \to \mathcal{O}$ defined as follows. If $M \in \mathcal{O}$ the vector space DM is $DM = \oplus_{\lambda \in \mathfrak{h}^*}(M_\lambda)^*$, the graded dual of M. We define the action of $\mathfrak{g}$ on DM by the formula

$$xf(m) = f(\tau(x)m), \ x \in \mathfrak{g}, f \in DM, m \in M,$$

where τ is the anti-involution on $\mathfrak{g}$ which fixes pointwise $\mathfrak{h}$ and takes the root space $\mathfrak{g}_\alpha$ to $\mathfrak{g}_{-\alpha}$ for all roots α.

Note that all weight spaces M_λ are finite dimensional. The above construction implies that $\dim(DM)_\lambda = \dim M_\lambda$. Hence D preserves the formal character of a module. As a consequence, $DL(\lambda) \simeq L(\lambda)$ for all $\lambda \in \mathfrak{h}^*$.

Let $\lambda \in \mathfrak{h}^*$. Then the *dual Verma module* with highest weight λ is

$$\nabla(\lambda) = D\Delta(\lambda). \tag{2.8}$$

Clearly, $\nabla(\lambda)$ has $L(\lambda)$ as its unique simple submodule. It has the same composition factors as $\Delta(\lambda)$.

We shall need the following easy and very useful result.

Proposition 2.6. *Let $\lambda, \mu \in X$. Then $\mathrm{Hom}_\mathfrak{g}(\Delta(\mu), \nabla(\mu)) = \delta_{\lambda,\mu}\mathbb{C}$ and $\mathrm{Ext}^1_\mathcal{O}(\Delta(\lambda), \nabla(\mu)) = 0$.*

Proof. The statement about the Hom-spaces is an easy weight argument. To check the Ext^1-vanishing suppose $0 \to \nabla(\mu) \to M \to \Delta(\lambda) \to 0$ is a short exact sequence. If $\mu \not> \lambda$ we have that λ is a highest weight of M. Hence the universal property of $\Delta(\lambda)$ gives a section of the surjection $M \to \Delta(\lambda)$, i.e. in this case the sequence splits. If $\mu > \lambda$ we dualize the sequence and find by the same argument a splitting of the surjection $DM \to \Delta(\mu)$. $\qquad\square$

Remark. In fact, $\mathrm{Ext}^i_\mathcal{O}(\Delta(\lambda), \nabla(\mu)) = 0$ for all $i > 0$, see e.g. [26], Theorem 6.12.

2.7. Projective modules in $\mathcal{O}$.

It is easy to see that the results above imply that $\Delta(\lambda)$ is a projective module in $\mathcal{O}$ whenever λ is dominant. Having $L(\lambda)$ as its unique simple quotient, it follows that $\Delta(\lambda)$ is, in this case, the projective cover of $L(\lambda)$. But then also $\Delta(\lambda) \otimes E$ is projective for any finite dimensional module E and this implies that for any weight μ we may find an appropriate E such that an indecomposable summand of $\Delta(\lambda) \otimes E$ maps onto $L(\mu)$. Thus $\mathcal{O}$ has enough projectives and moreover, all projective modules in $\mathcal{O}$ have Δ-filtrations (Verma flags). The last claim follows by the easy observation that (for any λ and any finite dimensional module E) the module $\Delta(\lambda) \otimes E$ has a Δ-filtration. If P has a Δ-filtration then we write $(P : \Delta(\mu))$ for the number of times $\Delta(\mu)$ occurs as a quotient in this filtration.

Denote by $P(\mu)$ a projective cover of $L(\mu)$. Then we have

Theorem 2.7. *[9]* $(P(\mu) : \Delta(\lambda)) = [\Delta(\lambda) : L(\mu)]$.

Proof. Since $\Delta(\lambda)$ and $\nabla(\lambda)$ have the same composition factors we see that the right hand side in the theorem equals the dimension of $\mathrm{Hom}_{\mathfrak{g}}(P(\mu), \nabla(\lambda))$. Now apply the functor $\mathrm{Hom}_{\mathfrak{g}}(-, \nabla(\mu))$ to a Δ-filtration of $P(\mu)$. Proposition 2.6 shows then that the same dimension equals the left hand side in the theorem. $\square$

Remark. The formula in Theorem 2.7 is often called *BGG reciprocity*. Bernstein, Gelfand, and Gelfand were inspired by a similar reciprocity for finite dimensional modular representations of Lie algebras (Humphreys reciprocity, cf. Section 5.4 below).

2.8. The Kazhdan-Lusztig conjecture. Soon the problem of determining the multiplicities $[\Delta(\lambda) : L(\mu)]$ (cf. Section 2.5) became the main open problem in this branch of representation theory. Verma had thought that he had an argument showing that all non-zero such multiplicities were 1, cf [43] (see also [26], Section 4.14). However, this quickly turned out to be wrong [8] and even though some multiplicities could be computed for low rank cases no one had any idea of a general formula, see footnote 5 in [44]. Then in 1979 Kazhdan and Lusztig made a major breakthrough [34]. They saw a connection to a new combinatorics on Coxeter groups, the now famous Kazhdan-Lusztig polynomials. These are themselves defined as certain base change coefficients (between the standard basis of Hecke algebras and a new "canonical" basis). They conjectured that the value at 1 of the polynomial attached to the pair (y, w) of elements of the finite Weyl group W would give the coefficient $(L(w \cdot \lambda) : \Delta(y \cdot \lambda))$ whenever $\lambda \in \mathfrak{h}^*$ is antidominant.

This conjecture, the Kazhdan-Lusztig conjecture for category $\mathcal{O}$, was proved in 1981 independently by Beilinson and Bernstein [7] and Brylinski and Kashiwara [11].

In modern terminology this result – when properly spiced up – says that a graded version of the category $\mathcal{O}$ is the categorification of the Hecke algebra $\mathcal{H}$ corresponding to W [41]. The basis in $K_0(\mathcal{O})$ consisting of graded Verma modules then corresponds to the standard basis in $\mathcal{H}$ while the counterparts of graded simple modules in $\mathcal{O}$ constitute the Kazhdan-Lusztig basis of $\mathcal{H}$.

3. The Bruhat Ordering on Coxeter Groups. Affine Weyl Groups

3.1. Bruhat cells.

Let G be a connected semisimple algebraic group over an algebraically closed field k. Let $T \leq G$ be a maximal torus and suppose the root system for (G, T) is R. Then the Weyl group W for R coincides with $W = N_G(T)/T$. For $w \in W$, we denote a representative of w in $N_G(T)$ by n_w, i.e. $w = n_w T$.

Fix a Borel subgroup $B \leq G$ containing T by demanding that the set of roots of B is $-R_+$. The double coset $B n_w B$ does not depend on the choice of representative $n_w \in N_G(T)$ for $w \in W$ and we shall denote it BwB. The Bruhat cell for w is then the image $C_w = BwB/B$ in the flag variety G/B. As a variety C_w is isomorphic to an affine space of dimension $l(w)$. Here $l : W \to \mathbb{N}$ denotes the length function on W. The cellular decomposition $G/B = \cup_{w \in W} C_w$ is called the Bruhat decomposition of G/B.

3.2. Schubert varieties and Bruhat ordering.

Let $w \in W$. The closure $X_w = \bar{C}_w$ of the Bruhat cell C_w is called a Schubert variety. In his famous "lost manuscript" [15] Chevalley defined an ordering $\leq$ on W by

Definition 3.1. Let $y, w \in W$. Then $y \leq w$ if and only if $X_y \subset X_w$

This ordering has become known as the Bruhat ordering although it would arguably have been more appropriate to call it the Chevalley ordering, see Borel's foreword in [15]. Chevalley proves in this paper that $y \leq w$ if and only if y may be obtained by deleting some of the factors in a reduced expression for w. This last condition makes sense for general Coxeter groups, and Verma seems to have been the first to explore this. He made detailed analysis of this ordering in [43] and [45]. In particular, he proved the exchange condition:

Proposition 3.2. *Let (W, S) be a Coxeter group. Suppose $t \in W$ is a reflection, i.e. t is a W-conjugate of an element of S. Suppose $w \in W$ satisfies $l(wt) < l(w)$. If $w = s_1 s_2 \cdots s_r$ for some $s_1, s_2, \cdots, s_r \in S$ then there exists $1 \leq i \leq r$ for which $wt = s_1 s_2 \cdots \widehat{s_i} \cdots s_r$.*

It should also be mentioned that Verma obtained an explicit formula for the Möbius function on an arbitrary Coxeter group [45]. Moreover, his conjecture on the so-called Weyl's dimension polynomial discussed in the first part of [44] involved key properties on Coxeter groups. This conjecture was solved by Hulsurkar [24], who followed up in later papers with several further results on Weyl groups and Coxeter groups. These are examples of topics we will not discuss any further in this article.

3.3. Affine Weyl groups. Let $E = \operatorname{span}_{\mathbb{R}} R$ denote the Euclidian space containing the root system R. If $\alpha \in R_+$ and $n \in \mathbb{Z}$ then we have an affine reflection $s_{\alpha,n} : E \to E$ defined by

$$s_{\alpha,n}\lambda = s_\alpha \lambda + n\alpha, \ \lambda \in E. \tag{3.1}$$

The group W_a generated by $\{s_{\alpha,n} \mid \alpha \in R_+, n \in \mathbb{Z}\}$ is the *affine Weyl group*. More precisely, it is the affine Weyl group of the dual root system $R^\vee$, see [10]. It is easily seen that W_a is the semi-direct product of W and the group of translations on E given by the root lattice $\mathbb{Z}R$. For instance, we have $s_{\alpha,n} = t_{n\alpha} \circ s_\alpha$ where for any $\lambda \in E$ we denote translation by λ by the symbol t_λ.

The affine Weyl group W_a is an example of an infinite Coxeter group. If we assume R to be indecomposable then W_a has a system of generators $\{s_0, s_1, \cdots, s_l\}$ where the $s_i = s_{\alpha_i}$ for $i > 0$ are the reflections with respect to the simple roots $\{\alpha_1, \cdots, \alpha_l\}$ in R_+. The remaining generator s_0 is $s_0 = s_{\alpha_0,1}$ where α_0 is the highest short root in R.

3.4. Affine Weyl groups and modular representations. Suppose the field k in Section 3.1 has characteristic $p > 0$. We then take a "p-version" of W_a which we denote W_p. It is the group generated by $\{s_{\alpha,np} \mid \alpha \in R_+, n \in \mathbb{Z}\}$. As an abstract group it is of course identical to W_a, but this version is better suited for use in representation theory of $G = G_k$.

We consider the "dot"-action of W_p on X. As fundamental domain for this action we may take the closure of the alcove

$$A_p = \{\lambda \mid 0 < \langle \lambda + \rho, \alpha^\vee \rangle < p \text{ for all } \alpha \in R_+\}. \tag{3.2}$$

Note that the generators (where now $s_0 = s_{\alpha_0,p}$) of W_p are the reflections of the walls of A_p (again assuming R is irreducible).

The philosophy of Verma is now that W_p should play a similar role in modular representation theory as W did in the representation theory discussed in Section 2. We shall describe this in some detail in the following sections.

4. Verma's Modular Conjectures

Throughout this section we will continue to use the notation G, T etc. from Section 3 but from now on the ground field k will always have characteristic $p > 0$.

4.1. Induced modules. Let $X = X(T)$ denote the character group of T. Then X is the integral lattice in $E = \operatorname{span}_{\mathbb{R}}(R)$ given by

$$X = \{\lambda \in E \mid \langle \lambda, \alpha^\vee \rangle \in \mathbb{Z} \text{ for all } \alpha \in R\}.$$

The dominant characters form the cone $X^+ = \{\lambda \in X \mid \langle \lambda, \alpha^\vee \rangle \geq 0$ for all $\alpha \in R_+\}$. We consider elements of X also as characters of B via the natural projection $B \to T$.

When M is a rational B-module we define the induced G-module

$$\text{Ind}_B^G M = \{f : G \to M \mid f(gb) = b^{-1}f(g), g \in G, b \in B\}.$$

In this definition it is understood that the maps in question are rational maps, i.e. maps of varieties (M being considered as the affine space of dimension $\dim_k M$). The G-action on $\text{Ind}_B^G M$ is given by $xf : g \mapsto f(x^{-1}g)$, $x, g \in G, f \in \text{Ind}_B^G M$.

Alternatively, we may view $\text{Ind}_B^G M$ as the module obtained as the set of global sections $H^0(M)$ for the vector bundle on G/B induced by M. More generally, we then have

$$R^i \text{Ind}_B^G M \simeq H^i(M), \quad \text{for all } i \geq 0.$$

Here the left hand side denotes the i-th right derived functor of Ind_B^G (as functor from the category of rational B-modules to the corresponding category for G) whereas the right hand side denotes the i-th coherent sheaf cohomology of the vector bundle on G/B induced by M. Recall that these modules are 0 for all $i > \dim G/B$ and are finite dimensional when M is.

4.2. Weyl modules. Let $\lambda \in X^+$. Then the Weyl module $\Delta(\lambda)$ with highest weight λ may be defined as the "reduction mod p" of the corresponding irreducible module for $G_\mathbb{C}$, the $\mathbb{C}$-version of G. More precisely, we choose a minimal $\mathbb{Z}$-lattice in this irreducible $G_\mathbb{C}$-module and tensor it by k. Then $\Delta(\lambda)$ is not necessarily irreducible for G but it has always a unique simple quotient which we denote $L(\lambda)$. In this way we get up to isomorphisms all finite dimensional irreducible G-modules.

Alternatively, we may consider $\nabla(\lambda) = \text{Ind}_B^G \lambda$, see Section 4.1. It is easy to see that this module has $L(\lambda)$ as its unique simple submodule. If w_0 denotes the longest element in W then we have $\nabla(\lambda) \simeq \Delta(-w_0\lambda)^*$. This fact is a consequence of Kempf's vanishing theorem [36] (see [2] and [22] for easy alternative proofs).

The name Weyl modules comes from [14] where these modules were studied carefully in the case $G = GL_n$, cf. also Section 4.6 below. Their characters are given by Weyl's famous character formula.

4.3. Linkage. As mentioned in Section 3 Verma's point of view in [44] was that the affine Weyl group W_p should play the same role for the category of finite dimensional modular representations of G as the ordinary Weyl group W does in the category $\mathcal{O}$, see Section 2. We explain this here and in the following sections.

Definition 4.1. We say that two weights $\lambda, \mu \in X$ are *linked* if they belong to the same W_p-orbit, i.e. if $\lambda = w \cdot \mu$ for some $w \in W_p$.

Let us write $\lambda \uparrow \mu$ if there exist $\alpha \in R_+, n \in \mathbb{Z}$ such that $\lambda = s_{\alpha,np} \cdot \mu$ and $\langle \mu + \rho, \alpha^\vee \rangle \geq np$. Note that the last condition means that $\mu - \lambda$ is a non-negative integer multiple of α. In particular $\lambda \leq \mu$ in the ordering on X induced by R_+.

Definition 4.2. We say that $\lambda \in X$ is *strongly linked* to $\mu \in X$ if there exists a sequence $\lambda = \lambda_0 \uparrow \lambda_1 \uparrow \cdots \uparrow \lambda_n = \mu$

In analogy with Corollary 2.5 we then have

Theorem 4.3. *(Strong linkage principle). Let* $\lambda, \mu \in X^+$. *If* $L(\mu)$ *is a composition factor of* $\Delta(\lambda)$ *then* μ *is strongly linked to* λ.

This result (or rather its consequence stated in Corollary 4.4 below) was conjectured by Verma in 1967 (cf. the first sentence of Section 5 in [44]). He suggested what he called a *"Harish-Chandra principle"* in characteristic p. J. E. Humphreys [25] introduced the term *"linkage"* and proved the *"infinitesimal linkage principle"* (cf. Section 5 below) for $p > h$ and applied it to representations of G. Later, J. C. Jantzen [28], [29] gave an alternative proof which extended the result to smaller values of p. Finally, the present author [1] generalised the statement in the theorem to the higher cohomology modules $H^i(\lambda)$ and proved it in this generality for all p. It should be emphasized that the inclusion of the higher cohomology - even though these modules are to this day poorly understood - is completely essential for the proof.

An important consequence of the strong linkage principle is

Corollary 4.4. *(Linkage principle) Let* M *be an indecomposable G-module. Then any two composition factors* $L(\mu)$ *and* $L(\lambda)$ *of* M *have linked highest weights, i.e.* $\mu \in W_p \cdot \lambda$.

Proof. We shall sketch the main steps in the deduction of this corollary from Theorem 4.3:

If M is a rational G-module we denote by $\mathrm{Ext}_G^i(M, -)$ the i'th right derived functor of the functor $\mathrm{Hom}_G(M, -)$ from the category of rational G-modules to the category of vector spaces. The corollary follows if we check

$$\mathrm{Ext}_G^1(L(\mu), L(\lambda)) = 0 \text{ unless } \mu \in W_p \cdot \lambda. \tag{4.1}$$

To see this we first observe that if a module M has no weights bigger than λ then $\mathrm{Ext}_G^1(M, \nabla(\lambda)) = 0$. In fact, suppose

$$0 \to \nabla(\lambda) \to E \to M \to 0$$

161

is a short exact sequence of G-modules. Then the assumption on M makes λ a highest weight of E and hence the universal property of the induced module $\nabla(\lambda)$ gives us a retraction of the inclusion $\nabla(\lambda) \to E$.

To check (4.1) we assume first that $\mu \not> \lambda$. Then the above argument implies that $\text{Ext}_G^1(L(\mu), \nabla(\lambda)) = 0$. The short exact sequence $0 \to L(\lambda) \to \nabla(\lambda) \to \nabla(\lambda)/L(\lambda) \to 0$ then gives $\text{Hom}_G(L(\mu), \nabla(\lambda)/L(\lambda)) \simeq \text{Ext}_G^1(L(\mu), L(\lambda))$. So Theorem 4.3 implies the corollary in this case.

If $\mu > \lambda$ we use duality to see that $\text{Ext}_G^1(L(\mu), L(\lambda)) \simeq \text{Ext}_G^1(L(\lambda), L(\mu))$. Now the same arguments as above with the roles of μ and λ reversed finish the proof. $\qquad\square$

For later use we record the following consequence of the linkage principle. We set

$$St_p = L((p-1)\rho). \qquad (4.2)$$

This is the *Steinberg module* which shows up in many crucial ways in modular representation theory.

Corollary 4.5. $St_p = \Delta((p-1)\rho)$.

Proof. There are no dominant weights strictly less than $(p-1)\rho$ which are linked to $(p-1)\rho$. Hence $\Delta((p-1)\rho)$ has St_p as its only composition factor. $\qquad\square$

4.4. Lusztig's conjecture. Shortly after the original Kazhdan-Lusztig conjecture [34] for Verma modules for complex semisimple Lie algebras Lusztig proposed an analogous conjecture [38] for Weyl modules in characteristic p. The idea is simply to replace the finite Weyl group by the affine Weyl group W_p, i.e. the composition factor multiplicity of a simple module in a Weyl module is conjectured to be the value at 1 of the ("inverse") Kazhdan-Lusztig polynomial whose indices are the pair of elements in W_p attached to the highest weights of the Weyl module and the simple module in question. Recall that W_p is the affine Weyl group for the dual root system (as predicted in [44] this is the Coxeter group that should govern the modular representations of G).

This conjecture is limited to the case where the highest weights belong to the "Jantzen region" (i.e. $\langle \lambda+\rho, \alpha^\vee \rangle \leq p(p-h+2)$ where h denotes the Coxeter number). By Steinberg's tensor product theorem [42] this is enough to obtain all irreducible characters when $p > 2h - 2$. It is known how to reduce this bound to $p > h - 1$, see e.g. the discussions in Section 8.22 of [31] and in the introduction to [6].

Lusztig's conjecture has been proved for large p in [6]. No explicit bound on p is given in loc. cit. This was recently remedied by Fiebig, [21] who obtained an explicit (but still very crude and huge) bound on p.

4.5. Independence of p. Verma's paper [44] ends with an epilogue with the subtitle *"a metamathematical (?) conjecture"*. This conjecture says that the modular representation theory for G should not depend on p but only on R (or rather W_a).

More precisely, let $\lambda \in X^+$ and suppose ν denotes the unique element in the closure of A_p intersected with $W_p \cdot \lambda$. If $\mu \in X^+ \cap W_p \cdot \nu$ then we pick y and w minimal in W_p such that $\lambda = w \cdot \nu$ and $\mu = y \cdot \lambda$. Verma's philosophy is that the number $[\Delta(\lambda) : L(\mu)]$ only depends on the the pair $y, w \in W_p$. As in Section 4.5 this assumes that λ is in the "Jantzen region". Alternatively, we should pass to the infinitesimal case (which we do in Section 5 below) where we can allow λ to be arbitrary in X.

We shall call the numbers $\{[\Delta(\lambda) : L(\mu)]\}_{\lambda,\mu}$ the combinatorics of the category of finite dimensional representations of G. Steinberg's tensor product theorem [42] reduces (for $p \geq 2h - 2$) the problem of describing this combinatorics to finding the finite part with λ, μ belonging to the Jantzen region. Hence (still for $p \geq 2h-2$) the Lusztig conjecture discussed in Section 4.5 shows that the combinatorics for G is determined by the Kazhdan-Lusztig polynomials for W_a, i.e. Lusztig's conjecture would verify Verma's metamathematical conjecture.

When quantum groups appeared on the scene in the mid 1980's the same philosophy could be extended to the combinatorics of their finite dimensional representations at roots of unity. If we consider a complex root of unity of order l then Lusztig formulated a conjecture in this quantum case [39] which is completely analogous to the modular case we have discussed. The idea is to replace p by l and describe the quantum combinatorics at such a root of 1 in terms of the Kazhdan-Lusztig polynomials attached to W_l. This conjecture was verified by building a bridge [34] (combined with [40] in the non-simply laced case) to the corresponding Kac-Moody algebras where the analogous Kazhdan-Lusztig conjecture was proved in [33]. It follows that in fact this quantum combinatorics is independent of l.

The proof in [6] of Lusztig's modular conjecture for large p was done by comparing the modular combinatorics with this quantum combinatorics. Thus we morally (i.e. for a given group G it is verified for almost all primes) proved Verma's metamathematical conjecture. On the other hand, Lusztig's conjecture does not give the full combinatorics when p is smaller than the Coxeter number and some features are definitely different for primes just slightly smaller than the Coxeter number, see e.g. [5], Example 7.9.

4.6. Homomorphisms between Weyl modules. With an eye on Verma's theorem (Theorem 2.1) and conjecture (Conjecture 2.4) for Verma modules it is natural to ask if there is a similar description for Weyl modules. It turns out that the situation here is much more complicated. Theorem 4.3

shows that if $\mathrm{Hom}_G(\Delta(\lambda), \Delta(\mu)) \neq 0$ then λ is strongly linked to μ. This was in fact first proved by Carter and Lusztig in the case $G = GL_n$, see [14]. The question of giving a condition on λ and μ which ensures the existence of a non-trivial homomorphism $\Delta(\lambda) \to \Delta(\mu)$ is still wide open. When $G = SL_2$ Carter and Cline solved the problem completely in [12] but in all other cases we seem far from a complete solution. Carter and Payne [13] have some results for SL_n in the case where $\mu - \lambda$ is a multiple of a positive root. In the general case this author has demonstrated that $\dim \mathrm{Hom}_G(\Delta(\lambda), \Delta(\mu)) = 1$ in the case when λ is maximal among all weights that are strongly linked but not equal to μ, see [3]. This result has been generalized to the case where λ is "close to" μ (in specific ways) by Koppinen, [37]. Note, however, that for a given λ there will often (always?) exist non-trivial homomorphisms from $\Delta(\lambda)$ to infinitely many $\Delta(\mu)$. For instance it is easy to see that $\mathrm{Hom}_G(\Delta(0), \Delta(2(p^r - 1)\rho)) = k$ for all $r \geq 0$ (observe that $\Delta(0) = k$).

Other related questions are:

i) Determine the dimension of $\mathrm{Hom}_G(\Delta(\lambda), \Delta(\mu))$

ii) Find the image of a given homomorphism $\Delta(\lambda) \to \Delta(\mu)$

iii) Determine $\mathrm{Ext}_G^i(\Delta(\lambda), \Delta(\mu))$

iv) Determine $\mathrm{Ext}_G^i(\nabla(\lambda), \Delta(\mu))$.

All these questions seem very hard. The answer to the first question is ≤ 1 in almost all known cases. However, a 2-dimensional Hom-space does occur for G of type D_{2n} in characteristic 2, see [16] but I see no reason why arbitrarily high dimensions couldn't be found in general. However, the mentioned exception is the only known example (to me) where the dimension is > 1.

In the second question it should be noted that homomorphisms between Weyl modules are seldom injective (in contrast with the Verma module case, see Section 2.2). It is therefore a very non-trivial problem to describe the image of such a homomorphism. Likewise, it is often hard to check whether the composite of two Weyl module homomorphisms is non-zero.

Note that if in question iv) the order of Δ and ∇ is reversed then the answer is $\delta_{i,0}\delta_{\mu,\lambda}$. This is seen by a generalisation of the arguments used in the proof of Corollary 4.4. Compare also with Proposition 2.6 and the remark following it.

5. Infinitesimal Case

As mentioned above the linkage principle was conjectured by Verma. Actually, in his formulations he dealt with the extended affine Weyl group (which contains all translations t_λ with $\lambda \in X$, not just $\lambda \in \mathbb{Z}R$). This was apparently due to the fact that he first gave an infinitesimal version of his conjecture. Historically, this infinitesimal case was also proved first (for $p > h$ by Humphreys [25]). The inspiration to both the conjecture and the proof of it by Humphreys came from the analogous situation in characteristic zero, cf Section 2.

In this section we shall turn things around and deduce the infinitesimal conjecture from the linkage principle in Section 4. Moreover, we shall formulate our results using infinitesimal subgroup schemes (Frobenius kernels) instead of dealing with the p-Lie algebra of G and its restricted universal enveloping algebra. For all the necessary details of our setup we refer to [31], which also contains generalisations to higher Frobenius kernels not discussed in this article.

5.1. The Frobenius homomorphism and its kernel. Let F : $G \to G$ be the Frobenius homomorphism on G coming from the p'th power map on k. If V is a rational G-module we get a new G-structure on V by composing the original one with F. We shall denote this new G-module by $V^{(1)}$ and call it the Frobenius twist of V.

The group scheme kernel of F is denoted G_1. Note that over any field - in particular over k - G_1 has just one point. That's why we call G_1 an infinitesimal algebraic group. Observe that G_1 acts trivially on any Frobenius twist $V^{(1)}$. Moreover, if M is a G-module whose restriction to G_1 is trivial then $M = N^{(1)}$ for some G-module N. In this case we sometimes write $N = M^{(-1)}$.

The representation theory of G_1 is "the same" as that of the p-Lie algebra $\mathfrak{g}_k$ of G. We may also obtain $\mathfrak{g}_k$ by first taking a Chevalley basis of the complex semisimple Lie algebra with root system R and then tensoring the $\mathbb{Z}$-span of this basis by k (assuming G simply connected). The representation theory of $\mathfrak{g}_k$ may also be thought of as the modules for the corresponding restricted enveloping algebra over k. This is the point of view in [44] and [25]. We shall prefer to study representations of G_1. In fact, we will enlarge G_1 and study representations of the group schemes $G_1T = F^{-1}(T)$ and $G_1B = F^{-1}(B)$. The advantage of including T in our group scheme is that we can then still decompose modules into weight spaces.

5.2. Infinitesimal modules. Let

$$X_p = \{\lambda \in X^+ \mid \langle \lambda, \alpha^\vee \rangle < p \text{ for all simple roots } \alpha\}.$$

We call this the set of restricted weights. A fundamental theorem first proved by Curtis [18] (in the language of $\mathfrak{g}_k$-modules) says

Theorem 5.1. If $\lambda \in X_p$ then $L(\lambda)$ restricted to G_1 remains irreducible.

Note that in this theorem we can obviously replace G_1 by either G_1T or G_1B. Moreover, for any $\mu \in X$, the Frobenius twist of the 1-dimensional T-

module (respectively, B-module) k_μ is clearly an irreducible G_1T-module (respectively, G_1B-module). Hence the modules of the form $L(\lambda)\otimes k_{p\mu}$ with $\mu \in X_p$ and $\mu \in X$ are irreducible G_1T and G_1B-modules.

Let $\lambda \in X$. We define

$$\widehat{\nabla}_1(\lambda) = \mathrm{Ind}_T^{G_1T} k_\lambda$$

and

$$\widetilde{\nabla}_1(\lambda) = \mathrm{Ind}_B^{G_1B} k_\lambda.$$

These are the *dual infinitesimal Verma modules*. Note that they are finite dimensional modules. It is easy to check that they have unique highest weight λ. In fact, a little closer examination gives

$$\mathrm{ch}\,\widehat{\nabla}_1(\lambda) = \mathrm{ch}\,\widetilde{\nabla}_1(\lambda) = \mathrm{ch}(St_p \otimes k_{\lambda-(p-1)\rho}), \tag{5.1}$$

where St_p is the Steinberg module from (4.2).

We set $\widehat{\Delta}_1(\lambda) = \widehat{\nabla}_1(2(p-1)\rho - \lambda)^*$ and $\widetilde{\Delta}_1(\lambda) = \widetilde{\nabla}_1(2(p-1)\rho - \lambda)^*$. These we call *infinitesimal Verma modules*. Note that by (5.1) we have $\mathrm{ch}\,\Delta_1(\lambda) = \mathrm{ch}\,\nabla_1(\lambda)$.

Standard arguments give that the infinitesimal Verma modules have a unique simple head while the dual infinitesimal Verma modules have simple socles. We denote by $\widehat{L}_1(\lambda)$, respectively $\widetilde{L}_1(\lambda)$ the head of $\widehat{\Delta}(\lambda)$, respectively of $\widetilde{\Delta}(\lambda)$.

Each weight in X has a p-adic expansion $\lambda = \lambda^0 + p\lambda^1$ for unique $\lambda^0 \in X_p, \lambda^1 \in X$. In the following upper index 0 and 1 on a weight will always refer to this expansion.

The above considerations show that for any $\lambda \in X$ we have

$$\widehat{L}_1(\lambda) \simeq L(\lambda^0) \otimes k_{p\lambda^1} \text{ as } G_1T\text{-modules and} \tag{5.2}$$

and

$$\widetilde{L}_1(\lambda) \simeq L(\lambda^0) \otimes k_{p\lambda^1} \text{ as } G_1B\text{-modules.} \tag{5.3}$$

Theorem 5.2. *(The infinitesimal strong linkage principle) Let $\lambda, \mu \in X$. If $\widehat{L}_1(\mu)$, resp. $\widetilde{L}_1(\mu)$, is a composition factor of $\widehat{\Delta}_1(\lambda)$, resp. $\widetilde{\Delta}_1(\lambda)$, then μ is strongly linked to λ.*

Proof. Observing that $\widehat{L}_1(\mu)$ is a G_1T-composition factor of $\widehat{\Delta}_1(\lambda)$ iff $\widetilde{L}_1(\mu)$ is a G_1B-composition factor of $\widetilde{\Delta}_1(\lambda)$, we see that it is enough to treat the G_1B-case. Moreover, $\widetilde{\Delta}_1(\lambda)$ has the same composition factors as $\widetilde{\nabla}_1(\lambda)$ so we shall switch to dual Verma modules. Finally, we note that $\widetilde{L}_1(\mu)$ is a composition factor of $\widetilde{\nabla}_1(\lambda)$ iff $\widetilde{L}_1(\mu + np\rho)$ is a composition factor of $\widetilde{\nabla}_1(\lambda + np\rho)$ and μ is strongly linked to λ iff $\mu + np\rho$ is strongly linked to $\lambda + np\rho$.

We shall say that λ is *"sufficiently dominant"* if all composition factors $\widetilde{L}_1(\mu)$ have $\mu \in X^+$. The above observations show that the theorem follows from Theorem 4.3 via the following proposition. $\qquad\square$

Proposition 5.3. *If λ is "sufficiently dominant" then we have for all $\mu \in X^+$*

$$[\widetilde{\nabla}_1(\lambda) : \widetilde{L}_1(\mu)] \leq [\nabla(\lambda) : L(\mu)].$$

Proof. Let $0 = \widetilde{F}_0 \subset \widetilde{F}_1 \subset \cdots \subset \widetilde{F}_r = \widetilde{\nabla}(\lambda)$ be a composition series of $\widetilde{\nabla}_1(\lambda)$. Applying the left exact functor $\mathrm{Ind}_{G_1B}^G$ to this we obtain the filtration $0 = F_0 \subset F_1 \subset \cdots \subset F_r = \mathrm{Ind}_{G_1B}^G \widetilde{\nabla}(\lambda) = \nabla(\lambda)$ where $F_i = \mathrm{Ind}_{G_1B}^G \widetilde{F}_i$ for all i and where the last equality comes from transitivity of induction functors. Our assumption means that the quotients in the $\widetilde{F}$-filtration have the form $\widetilde{L}_1(\mu)$ with $\mu \in X^+$. Note that then also $\mu^1 \in X^+$ and we get $R^i \mathrm{Ind}_{G_1B}^G \widetilde{L}_1(\mu) = R^i \mathrm{Ind}_{G_1B}^G (L(\mu^0) \otimes k_{p\mu^1}) = L(\mu^0) \otimes (R^i \mathrm{Ind}_B^G k_{\mu^1})^{(1)}$. Here the last equality uses the tensor identity for (derived) induction together with the fact that $R^i \mathrm{Ind}_{G_1B}^G (E^{(1)}) = (R^i \mathrm{Ind}_B^G(E))^{(1)}$. By Kempf's theorem [36] we therefore see that the quotients in the F-filtration are $L(\mu^0) \otimes \nabla(\mu^1)^{(1)}$. Noting that this quotient contains $L(\mu)$ as a composition factor, the inequality follows. $\qquad\square$

Remark. The arguments in the above proof yield the more general formula

$$[\nabla(\lambda) : L(\mu)] = \sum_{\nu} [\widetilde{\nabla}_1(\lambda) : \widetilde{L}_1(\mu^0 + p\nu)][\nabla(\nu) : L(\mu^1)]$$

for all "sufficiently dominant" λ and all $\mu \in X^+$. Using Euler character arguments based on Kempf's vanishing theorem, one obtains a formula valid for all $\lambda \in X^+$, see Theorem 3 in [4].

5.3. Projective G_1T-modules.

Clearly a G_1T-module is projective iff its dual M^* is injective. It is immediate that the category of finite dimensional G_1T-modules has enough injectives and projectives. In fact, the coordinate ring $k[G_1]$ is an injective G_1T-module because induction functors take injective modules to injective modules and by definition $k[G_1] = \mathrm{Ind}_T^{G_1T} k$. (Note that we have also $k[G_1] = \mathrm{Ind}_B^{G_1B} k$ but the trivial module k is not injective for B. In fact, there are no finite dimensional injective G_1B-modules and this is the reason we restrict our attention in this section to G_1T). If M is an arbitrary finite dimensional G_1T-module then M imbeds into the injective module $M \otimes k[G_1]$.

An alternative starting point for producing projective G_1T-modules is

Proposition 5.4. *St_p is a projective G_1T-module.*

167

Proof. By Corollary 4.5 we have $St_p = \Delta((p-1)\rho)$. Hence by Weyl's dimension formula we have

$$\dim St_p = \dim \Delta((p-1)\rho) = p^N,$$

where N denotes the number of positive roots. But we also have $\dim \widehat{\Delta}_1(\lambda) = p^N$ for all λ and since St_p is a quotient of $\widehat{\Delta}((p-1)\rho)$ we conclude that $St_p = \widehat{\Delta}_1((p-1)\rho) = \widehat{\nabla}_1((p-1)\rho)$. All $\widehat{\nabla}_1(\lambda)$ are injective for B_1^{opp} where B^{opp} is the opposite Borel subgroup (the one corresponding to the positive roots) so in particular we conclude that St_p is B_1^{opp}-injective. By symmetry it is also B_1-injective and hence G_1- (and thus $G_1 T$-) injective, see [31] 11.4. $\square$

For $\lambda \in X$ we denote by $\widehat{Q}_1(\lambda)$ the projective cover of $\widehat{L}_1(\lambda)$. By Proposition 5.4 we have $\widehat{Q}_1((p-1)\rho) = St_p$. We then get

Proposition 5.5. *Let* $\lambda \in X_p$. *Then* $\widehat{Q}_1(\lambda)$ *is uniquely a* $G_1 T$-*summand of* $St_p \otimes L((p-1)\rho + w_0\lambda)$

Proof. $\mathrm{Hom}_G(St_p \otimes L((p-1)\rho + w_0\lambda), L(\lambda)) = \mathrm{Hom}_G(St_p, L((p-1)\rho + w_0\lambda)^* \otimes L(\lambda)) = \mathrm{Hom}_G(St_p, L((p-1)\rho - \lambda) \otimes L(\lambda)) = k$ $\square$

Remark.

i) Clearly, we have for arbitrary $\lambda \in X$ an isomorphism $\widehat{Q}_1(\lambda) \simeq \widehat{Q}_1(\lambda^0) \otimes k_{p\lambda^1}$.

ii) It is tempting to conjecture that the $G_1 T$-summand of $St_p \otimes L((p-1)\rho + w_0\lambda)$ in this proposition is in fact a G-summand, see [27]. However, this has only been verified for $p \geq 2h - 2$, see [31]. For smaller p it is still an open problem whether $\widehat{Q}_1(\lambda)$ has a G-structure for all $\lambda \in X_p$. If λ is "close to" $(p-1)\rho$ (i.e. in the closure of one of the top alcoves in X_p) this is true.

iii) A *tilting module* for G is a module having both a Δ- and a ∇-filtration. The indecomposable tilting modules are parametrized by their highest weights [20]. If $\lambda \in X^+$ we denote by $T(\lambda)$ the indecomposable tilting module with highest weight λ. Then Donkin conjectures that for $\lambda \in X_p$ we have $\widehat{Q}_1(\lambda) \simeq T(2(p-1)\rho + w_0\lambda))$, [20]. It is easily seen from the above that $2(p-1)\rho + w_0\lambda$ is the unique highest weight of $\widehat{Q}_1(\lambda)$. Moreover, when $p \geq 2h - 2$ so that $\widehat{Q}_1(\lambda)$ has a G-structure it is easy to check that it is also tilting. In other words, Donkin's conjecture is true in that case. For smaller primes it has so far only been verified for type A_2 and B_2.

5.4. Humphreys' reciprocity law. As we noticed in the proof of Proposition 5.4 we have $St_p \simeq \widehat{\Delta}_1((p-1)\rho) \simeq \widehat{\nabla}_1((p-1)\rho)$. It follows that $St_p \otimes M$ has both a $\widehat{\Delta}$- and a $\widehat{\nabla}$-filtration (i.e. is G_1T-tilting) for any finite dimensional G_1T-module M. In particular, we see that therefore any projective G_1T-module has this property: For $\widehat{Q}_1(\lambda)$ with $\lambda \in X_p$ this follows from Proposition 5.5. It is then inherited by all $\widehat{Q}_1(\lambda)$, λ arbitrary via Remark i) in Section 5.3.

In analogy with previous notation when a G_1T-module M has a $\widehat{\Delta}_1$-filtration we write $(M : \widehat{\Delta}_1(\lambda))$ for the number of times $\widehat{\Delta}_1(\lambda)$ occurs as quotient in such a filtration. We have [25]

Theorem 5.6. *(Humphreys' reciprocity)* Let $\lambda, \mu \in X$. Then $(\widehat{Q}_1(\lambda) : \widehat{\Delta}_1(\mu)) = [\widehat{\Delta}_1(\mu) : \widehat{L}_1(\mu)]$.

Proof. Completely analogous to the proof we gave of Theorem 2.7. □

References

[1]   H. H. Andersen, The strong linkage principle. J. Reine Angew. Math. **315** (1980), 53–59.

[2]   H. H. Andersen, The Frobenius morphism on the cohomology of homogeneous vector bundles on G/B. Ann. Math. (2) **112** (1980), no. 1, 113–121.

[3]   H. H. Andersen, On the structure of the cohomology of line bundles on G/B. J. Algebra **71** (1981), no. 1, 245–258.

[4]   H. H. Andersen, Modular representations of algebraic groups. The Arcata Conference on Representations of Finite Groups (Arcata, Calif., 1986), 23–36, Proc. Sympos. Pure Math., **47**, Part 1, Amer. Math. Soc., Providence, RI, 1987.

[5]   H. H. Andersen, Finite-dimensional representations of quantum groups. Algebraic groups and their generalizations: quantum and infinite-dimensional methods (University Park, PA, 1991), 1–18, Proc. Sympos. Pure Math., **56**, Part 2, Amer. Math. Soc., Providence, RI, 1994.

[6]   H. H. Andersen, J. C. Jantzen and W. Soergel, Representations of quantum groups at a $p$pth root of unity and of semisimple groups in characteristic p: independence of p, Astérisque No. **220** (1994).

[7]   A. Beilinson, J. Bernstein; Localisation de $\mathfrak{g}$-modules. C. R. Acad. Sci. Paris Sér I Math. **292** (1981), no. 1, 15–18.

[8]   I. Bernstein, I. Gelfand, S. Gelfand, Structure of representations that are generated by vectors of highest weight. (Russian) Funckcional. Anal. i Prilozen. 5 (1971), no. 1, 1–9.

[9]   I. Bernstein, I. Gelfand, S. Gelfand; A certain category of $\mathfrak{g}$-modules. Funkcional. Anal. i Prilozen. **10** (1976), no. 2, 1–8.

[10] N. Bourbaki, Groupes et algèbres de Lie. Chapitres 4, 5 et 6. Masson, Paris, 1968.

[11] J.-L., Brylinski, M. Kashiwara; Kazhdan-Lusztig conjecture and holonomic systems. Invent. Math. **64** (1981), no. 3, 387–410.

[12] R. W. Carter and E. Cline, The submodule structure of Weyl modules for groups of type A_1. Proceedings of the Conference on Finite Groups (Univ. Utah, Park City, Utah, 1975), pp. 303–311. Academic Press, New York, 1976

[13] R. W. Carter and M. T. J. Payne, On homomorphisms between Weyl modules and Specht modules. Math. Proc. Cambridge Philos. Soc. **87** (1980), no. 3, 419–425.

[14] R. W. Carter and G. Lusztig, On the modular representations of the general linear and symmetric groups. Math. Z. 136 (1974), 193–242.

[15] C. Chevalley, Sur les décompositions cellulaires des espaces G/B. With a foreword by Armand Borel. Proc. Sympos. Pure Math., 56, Part 1, Algebraic groups and their generalizations: classical methods (University Park, PA, 1991), 1–23, Amer. Math. Soc., Providence, RI, 1994.

[16] E. Cline, B. Parshall, L. Scott; Cohomology of finite groups of Lie type. I. Inst. Hautes Études Sci. Publ. Math. No. **45** (1975), 169–191.

[17] E. Cline, B. Parshall, L. Scott; Finite-dimensional algebras and highest weight categories. J. Reine Angew. Math. **391** (1988), 85–99.

[18] C.W. Curtis, Representations of Lie algebras of classical type with applications to linear groups. J. Math. Mech.**9** (1960) 307–326.

[19] J. Dixmier; Enveloping algebras. Revised reprint of the 1977 translation. Graduate Studies in Mathematics, **11**. American Mathematical Society, Providence, RI, 1996.

[20] S. Donkin, On tilting modules for algebraic groups. Math. Z. **212** (1993), no. 1, 39–60

[21] P. Fiebig, Lusztig's conjecture as a moment graph problem. Bull. Lond. Math. Soc. **42** (2010), no. 6, 957–972.

[22] W. J. Haboush, A short proof of the Kempf vanishing theorem. Invent. Math. **56** (1980), no. 2, 109–112.

[23] Harish-Chandra, On some applications of the universal enveloping algebra of a semisimple Lie algebra, Trans. Amer. Math. Soc **70** (1951), 28–96.

[24] S. G. Hulsurkar, Proof of Verma's conjecture of Weyl's dimension polynomial. Invent. Math. **27** (1974), 45–52.

[25] J.E. Humphreys; Modular representations of classical Lie algebras and semisimple groups. J. Algebra **19** (1971), 51–79.

[26] J.E. Humphreys; Representations of semisimple Lie algebras in the BGG category $\mathcal{O}$. Graduate Studies in Math. **94**. American Mathematical Society, Providence, RI, 2008.

[27] J. E. Humphreys and D.-n. Verma, Projective modules for finite Chevalley groups. Bull. Amer. Math. Soc. 79 (1973), 467–468.

[28] J. C. Jantzen, Zur Charakterformel gewisser Darstellungen halbeinfacher Gruppen und Lie-Algebren. Math. Z. 140 (1974), 127–149.

[29] J. C. Jantzen, Darstellungen halbeinfacher Gruppen und kontravariante Formen. J. Reine Angew. Math. 290 (1977), 117–141.

[30] J. C. Jantzen, Moduln mit einem höchsten Gewicht. Lecture Notes in Mathematics, 750. Springer, Berlin, 1979.

[31] J. C. Jantzen, Representations of algebraic groups. Second edition. Mathematical Surveys and Monographs, bf 107. American Mathematical Society, Providence, RI, 2003.

[32] M. Kashiwara, T. Tanisaki; Kazhdan-Lusztig conjecture for affine Lie algebras with negative level. Duke Math. J. 77 (1995), 21–62.

[33] M. Kashiwara, T. Tanisaki; Kazhdan-Lusztig conjecture for affine Lie algebras with negative level. II. Nonintegral case. Duke Math. J. 84 (1996), 771–813.

[34] D. Kazhdan, G. Lusztig; Representations of Coxeter groups and Hecke algebras. Invent. Math. 53 (1979), no. 2, 165–184.

[35] D. Kazhdan, G. Lusztig; Tensor structures arising from affine Lie algebras I-IV. J. Amer. Math. Soc. 6, 905–947, 949–1011 and 7 (1994), 335–381, 383–453.

[36] G. R. Kempf, Linear systems on homogeneous spaces. Ann. Math. (2) 103 (1976), no. 3, 557–591.

[37] M. Koppinen, Homomorphisms between neighbouring Weyl modules. J. Algebra 103 (1986), no. 1, 302–319.

[38] G. Lusztig, Some problems in the representation theory of finite Chevalley groups. The Santa Cruz Conference on Finite Groups (Univ. California, Santa Cruz, Calif., 1979), pp. 313–317, Proc. Sympos. Pure Math., 37, Amer. Math. Soc., Providence, R.I., 1980

[39] G. Lusztig, Modular representations and quantum groups. Classical groups and related topics (Beijing, 1987), 59–77, Contemp. Math.,82, Amer. Math. Soc., Providence, RI, 1989.

[40] G. Lusztig, Monodromic systems on affine flag manifolds, Proc. Roy. Soc. London Series A 445 (1994), 231–246.

[41] W. Soergel, Kategorie O, perverse Garben und Moduln über den Koinvarianten zur Weylgruppe. J. Amer. Math. Soc. 3 (1990), no. 2, 421–445.

[42] R. Steinberg, Representations of algebraic groups. Nagoya Math. J.22 (1963) 33–56

[43] D.-n. Verma, Structure of certain induced representations of complex semisimple Lie algebras. Bull. Amer. Math. Soc. 74 1968 160–166.

171

[44] D.-n. Verma, The rôle of affine Weyl groups in the representation theory of algebraic Chevalley groups and their Lie algebras. Lie groups and their representations (Proc. Summer School, Bolyai János Math. Soc., Budapest, 1971), pp. 653–705. Halsted, New York, 1975.

[45] D.-n.Verma, Möbius inversion for the Bruhat ordering on a Weyl group. Ann. Sci. École Norm. Sup. (4) 4 (1971), 393–398.

172

The Work of Wiener and Masani on Prediction Theory and Harmonic Analysis

V. Mandrekar*

Introduction

The work of Norbert Wiener and Pesi Masani started at Indian Statistical Institute in Kolkata during the visit of Wiener in 1955-56. Wiener [18] had done substantial work on the prediction problem for univariate (weakly) stationary (henceforth, stationary) processes and had partial results for multivariate stationary processes. A.N. Kolmogorov [7] had studied the univariate case in detail, with emphasis on a fundamental theorem of H. Wold. In addition, V. Zasuhin [20] had announced partial results. Because of the connection between prediction theory (of interest to Wiener) and factorization of matrix-valued functions (of interest to Wiener [18] p.150 and Masani [12]), their collaboration produced results and techniques which had a lasting influence on prediction theory, analysis and operator theory (Nagy and Foias, see [13]).

In the next section, we start by describing the concepts involved in prediction theory in the univariate case. The material is based on [9].

Finally, it will be incomplete to give the influence of Masani in Mathematics without mentioning his mentoring of students as described in [4]. This led to these students contributing to Mathematics in general.

1. Motivation for Wiener-Masani Work

In order to understand the significance and motivation for the fundamental work of Wiener-Masani, it is necessary to describe the work of Kolmogorov and Wiener on the prediction problem for stationary processes. The rigorous work of Kolmogorov is motivated from that of Wold.

Let $(\Omega, \mathcal{F}, P)$ be a probability space and $L^2(\Omega, \mathcal{F}, P)$ be the space of complex-valued square integrable functions on $(\Omega, \mathcal{F}, P)$. We say that $\{X_n, n \in$

---

*Department of Statistics and Probability, Michigan State University.

$\mathbb{Z}\} \subseteq L^2(\Omega, \mathcal{F}, P)$ is a stationary process if $EX_n = c$, a constant, and $EX_n\overline{X}_m$ for $n, m \in \mathbb{Z}$ is given by a function r on $\mathbb{Z}$ with

$$r(n - m) = EX_n\overline{X}_m.$$

Without loss of generality, we can assume $c = 0$. Let us associate with a stationary process, subspaces $L(X) = \overline{\mathrm{sp}}\{X_n, n \in \mathbb{Z}\}$, $L(X : n) = \overline{\mathrm{sp}}\{X_m, m \leq n\}$ for $n \in \mathbb{Z}$ (called past and present up to n). The object of the linear prediction problem is to predict X_n with $u \in L(X : n - 1)$ such that $E(X_n - u)^2$ is minimized over all $u \in L(X : n - 1)$. If we denote by P_M the projection onto a subspace M of $L(X)$, then clearly the predictor is $\hat{X}_n = P_{L(X:n-1)}X_n$, and the error is $E(X_n - \hat{X}_n)^2$ which is the variance of the innovation $\nu_n = X_n - \hat{X}_n$.

Wold's work shows that every stationary process can be written as the sum of two orthogonal stationary processes, one having 0 innovation and the other non-zero.

We now derive the Wold decomposition. We denote by $L(X : -\infty) = \bigcap_{n \in \mathbb{Z}} L(X : n)$, the remote past, and $W_n = L(X : n) \ominus L(X : n - 1)$ for each n. Then it is easy to see

Lemma 1.1. a) $L(X : n) = \sum_{k=0}^{\infty} \oplus W_{n-k} \oplus L(X : -\infty)$ and

b) $L(X) = \sum_{k=-\infty}^{+\infty} \oplus W_k \oplus L(X : -\infty)$.

Define $UX_n = X_{n+1}$, $n \in \mathbb{Z}$ and extend it to a unitary operator on $L(X) \to L(X)$, again calling it U. Then using the fact $UL(X : n) = L(X : n + 1)$, we get

$$P_{L(X:n+1)}U = UP_{L(X:n)},$$

$\nu_{n+1} = U\nu_n$ and $UL(X : -\infty) = L(X : -\infty)$. Using these, we get

Theorem 1.1 (Wold Decomposition). Let $\{X_n, n \in \mathbb{Z}\}$ be a stationary process. Then $X_n = X_n^{(1)} + X_n^{(2)}$ with $L(X^{(1)}) \perp L(X^{(2)})$, $L(X^{(1)} : -\infty) = \{0\}$, and $L(X^{(2)} : n) = L(X^{(2)} : -\infty)$ for all n. In addition, $L(X^{(j)}) \subseteq L(X)$ $(j = 1, 2)$, the decomposition is unique and $\{X_n^{(j)}, n \in \mathbb{Z}\}$ are stationary.

We call $\mathbf{X} = \{X_n, n \in \mathbb{Z}\}$ purely non-deterministic (regular) provided $L(X : -\infty) = \{0\}$ and deterministic (singular) if $L(X : -\infty) = L(X : n)$ for all n. Now if X_n is regular, we write

$$X_0 = \sum_{k=-\infty}^{0} a_k\nu_k$$

$$X_n = \sum_{k=-\infty}^{0} a_kU^n\nu_k = \sum_{k=-\infty}^{0} a_k\nu_{n+k} = \sum_{k'=-\infty}^{n} a_{k'-n}\nu_{k'}.$$

A major part of Kolmogorov-Wiener work ([7], [18]) for a univariate stationary process is to get analytic conditions for the process to be regular or singular. This is done by using Bochner's Theorem to obtain

$$r(n) = \int_{-\pi}^{\pi} e^{in\lambda} dF(\lambda).$$

We denote the spectral measure of $\mathbf{X}$ by $F(\lambda)$, and the density of F with respect to Lebesgue measure σ by $f(\lambda)$. The space $L^2(F) = L^2((-\pi, \pi], \mathcal{B}(-\pi, \pi], F)$ is called the spectral domain. The map $V : X_n \to e^{in\cdot}$ can be extended to a unitary operator from the time domain $L(X)$ onto the spectral domain $L^2(F)$. One then gets that $\{X_n\}$ is purely non-deterministic if and only if $f(\lambda) = |\varphi(\lambda)|^2$ where $\hat{\varphi}(k) = 0$ for $k < 0$. If $X_n = \sum\limits_{k=0}^{\infty} a_k \nu_{n-k}$, then $\varphi(\lambda) = \sum\limits_{k=0}^{\infty} \bar{a}_k e^{ik\lambda}$.

If $\mathbf{X}$ is singular, then F is singular with respect to Lebesgue measure. Thus the Wold decomposition is equivalent to the Lebesgue decomposition of F. Let us consider the map from $L^2(\mathbb{T})$ to $H^2(\mathbb{T}) = \{\varphi \in L^2(\mathbb{T}) : \hat{\varphi}(k) = 0, k < 0\}$ as

$$\varphi = \sum_{n=-\infty}^{\infty} \hat{\varphi}(n) e^{in\lambda} \longmapsto (\varphi)_+ = \sum_{n=0}^{\infty} \hat{\varphi}(n) e^{in\lambda}.$$

We observe that $\varphi \in H^2(\mathbb{T})$ does not vanish on a set of positive Lebesgue measure without being identically zero. Now if $f(\lambda) = |\varphi(\lambda)|^2$, $\varphi \in H^2(\mathbb{T})$, then we get that

$$X_n = V^{-1}(e^{in\cdot} \varphi(\cdot))$$

gives a stationary process with spectral density f and $\xi_k = V^{-1}(e^{ik\cdot})$ giving

$$X_n = \sum_{k=-\infty}^{0} \hat{\varphi}(k) \xi_{n-k}^{\varphi}.$$

Thus $L(X : n) \subseteq L(\xi : n)$. We want to know for what choice of φ, we can get

$$L(X : n) = L(\xi : n), \quad n \in \mathbb{Z}.$$

Let us write the representation for X corresponding to each φ factorization by

$$X_n = \sum_{k=0}^{\infty} \hat{\varphi}(n-k) \xi_k^{\varphi}.$$

Then we want a φ such that $L(\xi^\varphi : n)$ is the smallest, as $X_1 - P_{L(\xi^\varphi:0)}X_1 = \hat{\varphi}(0)$. Hence we get a factorization corresponding to this φ satisfies $|\hat{\varphi}(0)|^2 \geq |\hat{\bar{\varphi}}(0)|^2$ for any other factorization $f(\lambda) = |\bar{\varphi}(\lambda)|^2$. Such a φ is called maximal. We observe that by a theorem of G. Szegö [17], one gets $f(\lambda) = |\varphi(\lambda)|^2$, $\varphi \in H^2(T)$ for $f \in L^1(T)$ if and only if $\int_0^{2\pi} \log f(\lambda)d\sigma > -\infty$.

Another analytic result one can prove is

$$\exp\left\{\int_0^{2\pi} \log f(\lambda)d\sigma\right\} = \inf_\psi \left[\int_0^{2\pi} e^{\psi(\lambda)} f(\lambda)d\sigma\right],$$

where inf is taken over all (real) functions ψ, such that $\int_0^{2\pi} \psi(\lambda)d\sigma = 0$ and f is non-negative with $\log f \in L^1(T)$. We indicate a proof of this. Note

$$\exp\left\{\int_0^{2\pi} \log f(\lambda)d\sigma\right\} = \exp\left\{\int_0^{2\pi} \log\left[e^{\psi(\lambda)} f(\lambda)\right]d\sigma\right\}$$

$$\leq \int_0^{2\pi} e^{\psi(\lambda)} f(\lambda)d\sigma, \text{ by Jensen inequality.}$$

Here equality holds for $\psi(\lambda) = \int_0^{2\pi} \log f(\lambda)d\sigma - \log f(\lambda)$.

Using approximation by real trigonometric polynomials, we can take ψ to be the real trigonometric polynomial $\psi(\lambda) = P(e^{i\lambda}) + P(e^{-i\lambda})$ where $P(z)$ is a polynomial in z with constant term zero. Then $e^{\psi(\lambda)} = |\exp(P(e^{i\lambda}))|^2$. Let $Q(z) = e^{P(z)} - 1$; since $1 + Q(z)$ is analytic and $Q(0) = 0$, we get

$$\exp\left\{\int_0^{2\pi} \log f(\lambda)d\sigma\right\} \geq \inf_Q \int_0^{2\pi} |1 + Q(e^{-i\lambda})|^2 f(\lambda)d\sigma.$$

Note that such Q's can be approximated by trigonometric polynomials $\tilde{P}(e^{-it})$, $\tilde{P}(0) = 0$. Thus,

$$\exp\left\{\int_0^{2\pi} \log f(\lambda)d\sigma\right\} \geq \inf_{\tilde{P}} \int_0^{2\pi} |1 + \tilde{P}(e^{-i\lambda})|^2 f(\lambda)d\sigma.$$

Now the right-hand side of the above equals, using the map V, $\sigma^2 = E|X_0 - P_{L(X:-1)}X_0|^2$, the prediction error. Since Szegö gives

$$\varphi(z) = \exp\left\{\frac{1}{2}\int_0^{2\pi} \log f(\theta)\frac{e^{i\theta} + z}{e^{i\theta} - z}d\sigma\right\},$$

176

we get $|\varphi(0)|^2 = E|X_0 - P_{L(\xi^\varphi:-1)}X_0|^2$. As in general $L(X:-1) \subseteq L(\xi^\varphi:-1)$, we get

$$\sigma^2 \geq |\varphi(0)|^2$$

and if φ is maximal, one gets $\sigma^2 = |\varphi(0)|^2$. Thus we get the result of Kolmogorov [8].

Theorem K1. Let $\{X_n, n \in \mathbb{Z}\}$ be a stationary process, then

a) $\sigma^2 = \exp\left\{\int_0^{2\pi} \log f_a(\theta)d\sigma\right\}$ where f_a is the density of spectral measure F with respect to σ

b) X is deterministic if and only if $\int_0^{2\pi} \log f_a(\theta)d\sigma = -\infty$.

In order to solve the prediction problem, one wants to express ν_n, the innovations, in terms of $\{X_n, n \in \mathbb{Z}\}$. That is done in the next result.

Theorem K2. Let $\{X_n, n \in \mathbb{Z}\}$ be a stationary purely non-deterministic process. Let f be its density and assume $f \in L^\infty(d\sigma)$. Then

$$\nu_n = \sum_{k=0}^{\infty} d_k X_{n-k} \quad \text{with} \quad \sum_k d_k^2 < \infty$$

if and only if

$$f^{-1} \in L^1(\sigma)$$

and $d_k = \int_0^{2\pi} e^{-ik\lambda} \frac{1}{\varphi(\lambda)} d\lambda$ where φ is maximal factor.

As one can see from the review of the Kolmogorov-Wiener work in the univariate case, solving of the problem requires combinations of methods from analytic function theory, geometry of time domain of stationary stochastic processes, isometry between the time and spectral domain. It turns out that, in the multivariate case studied by Wiener and Masani, the time domain is a Hilbert module and the spectral domain involves square integrable matrix-valued functions with respect to a non-negative definite matrix-valued measure. Lebesgue decomposition of matrix-valued measure is not obvious, and factorization of matrix-valued functions in $L^1(\sigma)$ in terms of $H^2(T)$ matrix-valued functions requires an extension of measure theory and analytic function theory. Even in the formulation of Szegö's theorem, the expression for the prediction error given in Theorem K1(a) is not easy. Wiener and Masani took up this formidable task and, as we shall see, they succeeded in overcoming these difficulties. In addition, they partially prove an analogue of Theorem K2. We present their work in slight generality following ideas of Masani [11].

2. Multivariate Stochastic Processes

We begin with the notation. Let H be a separable Hilbert space and H^q be Cartesian product of q-dimensional (column) vectors with components in H. We endow it with a matrix "inner product," called the Gram-matricial structure, as follows: For $\mathbf{f}, \mathbf{g} \in H^q$,

$$((\mathbf{f}, \mathbf{g})) = [(f_i, g_j)], \quad \mathbf{f} = (f^{(i)})_{i=1}^q, \quad \mathbf{g} = (g^{(j)})_{j=1}^q,$$

with $[(\ )]$ being a matrix. We denote

$$\mathbf{f} \perp \mathbf{g} \text{ if and only if } ((\mathbf{f}, \mathbf{g})) = \mathbf{0}$$

A complex-valued inner product $(\mathbf{f}, \mathbf{g}) = \text{trace}((\mathbf{f}, \mathbf{g}))$ and norm $|\mathbf{f}|_E = \sqrt{\text{trace}(\mathbf{f}, \mathbf{f})}$ provide a topology on H^q. We take linear combinations of vectors $\mathbf{f}_k$ in H^q with $q \times q$ matrix-valued coefficients. Using this, we can define linear manifolds and (closed) subspaces of H^q. We get

(2.1) $\mathcal{M}$ is a subspace of H^q if and only if $\mathcal{M} = M^q$ where M is a subspace of H generated by $\{f^{(i)} : \mathbf{f} \in \mathcal{M}, i = 1, 2, \ldots, q\}$.

We denote by $\overline{\text{sp}}(\mathbf{f})$, $\overline{\text{sp}}\{\mathbf{f}_j, j \in J\}$ subspaces spanned by $\mathbf{f}$ and the family $\{\mathbf{f}_j, j \in J\}$. One can easily show that $\mathbf{f} \in H^q$ has a unique projection $(\mathbf{f}|\mathcal{M})$ onto a subspace $\mathcal{M}$ such that $(\mathbf{f}|\mathcal{M}) = (f^{(i)}|M)_{i=1}^q$, where $f^{(i)}|M$ denotes the projection of $f^{(i)}$ onto the subspace M, $i = 1, 2, \ldots, q$. Clearly $(\mathbf{f}|\mathcal{M}) \in \mathcal{M}$ and $\mathbf{f} - (\mathbf{f}|\mathcal{M}) \perp \mathcal{M}$.

It is easy to see that $\mathcal{M} = \overline{\text{sp}}(\mathbf{f})$ if and only if $\mathbf{f} \in \mathcal{M}$ and $\mathbf{0} \neq \mathbf{g} \in \mathcal{M}$ imply $\mathbf{g} \not\perp \mathbf{f}$. In [19], $H = L^2(\Omega, \mathcal{F}, P)$ and the above conclusions are proved. We begin now with time domain analysis. We shall call $\{\mathbf{f}_n, n \in \mathbb{Z}\}$ a weakly stationary process if $\mathbf{f}_n \in H^q$ and $((\mathbf{f}_n, \mathbf{f}_m)) = R(m - n)$, the covariance matrix-valued function R on $\mathbb{Z}$, depends only on the difference.

As in the univariate case, let us define $\mathcal{M}_n = \mathcal{M}(\mathbf{f} : n) = \overline{\text{sp}}\{\mathbf{f}_k, k \leq n\}$, $\mathcal{M}_{-\infty} = \bigcap_{n=-\infty}^\infty \mathcal{M}_n$. We know that $U\mathbf{f}_n^i = \mathbf{f}_{n+1}^i$ defines a unitary operator on H and we can write

$$U(\mathbf{f}_n) = (U\mathbf{f}_n^i)_{i=1}^q = \mathbf{f}_{n+1}.$$

Using a similar argument $UP_{\mathcal{M}_n} = P_{\mathcal{M}_{n+1}}U$ where $P_{\mathcal{M}_n}\mathbf{f} = (\mathbf{f}|\mathcal{M}_n)$ for $\mathbf{f}$ in H^q. Then we get $U\mathcal{M}_n = \mathcal{M}_{n+1}U$ and hence, just as in the univariate case, we get U acting as a projection on $\mathcal{M}_n$ equals the projection of $\mathcal{M}_{n+1}$ acting on U. Define $\mathbf{g}_n = \mathbf{f}_n - (\mathbf{f}_n|\mathcal{M}_{n-1})$ and $\mathcal{W}_n = \overline{\text{sp}}\{\mathbf{g}_n\}$ for each n. Then one gets with $\mathcal{M}_\infty = \overline{\text{sp}}\{\mathbf{f}_n, n \in \mathbb{Z}\}$

(2.2) a) $\mathcal{M}_n = \sum_{k=0}^{\infty} \oplus \mathcal{W}_{n-k} \oplus \mathcal{M}_{-\infty}$

b) $\mathcal{M}_\infty = \sum_{k=-\infty}^{+\infty} \oplus \mathcal{W}_k \oplus \mathcal{M}_{-\infty}$

As in the one-dimensional case, we call [19] a process purely non-deterministic if $\mathcal{M}_{-\infty} = \{0\}$ and deterministic if $\mathcal{M}_n = \mathcal{M}_{-\infty}$. Using this, Wiener-Masani prove the Wold Decomposition for a stationary process $\{\mathbf{f}_n, n \in \mathbb{Z}\}$.

Wold Decomposition ([19] Thm. 6.11). Let $G = ((\mathbf{g}_0, \mathbf{g}_0))$. Then

$$\mathbf{f}_n = \sum_{k=0}^{\infty} A_k \mathbf{g}_{n-k} + (\mathbf{f}_n | \mathcal{M}_{-\infty}), \quad \mathbf{g}_j \perp (\mathbf{f}_n | \mathcal{M}_{-\infty}),$$

where $A_k G = ((\mathbf{f}_0, \mathbf{g}_{-k}))$, $A_0 \mathbf{g}_0 = \mathbf{g}_0$ and $\sum_{k=0}^{\infty} |A_k \sqrt{G}|_E^2 < \infty$, $A_0 \sqrt{G} = \sqrt{G}$. Note that A_k are not necessarily unique, but $A_k G$ and $A_k \sqrt{G}$ are unique.

The above result gives the Wold Decomposition in terms of the purely non-deterministic part $\left(\sum_{k=0}^{\infty} A_k \mathbf{g}_{n-k} \right)$ and the deterministic part $((\mathbf{f}_n | \mathcal{M}_{-\infty}))$ and the moving average representation of the purely non-deterministic part. The process $\{\mathbf{g}_n, n \in \mathbb{Z}\}$ is stationary and is the innovation process of $\{\mathbf{f}_n, n \in \mathbb{Z}\}$. Earlier, Doob [2] has given this result under full-rank assumption, i.e. G being invertible. The following theorem is a generalization of Kolmogorov's Theorem ([7], Thm. 19).

The equivalence of the following conditions (a) and (b) can be easily derived from the above theorem by using $\varphi_n = \mathbf{g}_n$ and $K = G$, and by using orthogonality of $\{\varphi_n\}$ and the shift $U\varphi_n = \varphi_{n+1}$ and $\mathcal{M}_n^f \subseteq \mathcal{M}_n^\varphi$. Kolmogorov defines **regularity** of a stationary process by requiring $(\mathbf{f}_0 | \mathcal{M}_n^f) \to 0$ as $n \to \infty$, which easily follows as $\mathcal{M}_n^f \subseteq \mathcal{M}_n^\varphi$.

Kolmogorov Theorem. Each of the following conditions is equivalent to regularity of a stationary stochastic process $\{\mathbf{f}_n\}_{-\infty}^{\infty}$:

a) $\mathbf{f}_n = \sum_{k=0}^{\infty} A_k \varphi_{n-k}$, $(\varphi_n, \varphi_m) = \delta_{m,n} K$, where K is a constant matrix,

b) $\mathcal{M}_{-\infty} = \{0\}$.

179

3. Spectral Analysis of Multivariate Processes

We note that for x belonging to $\mathcal{M}_\infty$,

$$\boldsymbol{U}\boldsymbol{x} = \begin{pmatrix} Ux^{(1)} \\ \vdots \\ Ux^{(q)} \end{pmatrix} = \left(\int_0^{2\pi} e^{-i\theta} \mathrm{d}E(\theta) x^{(i)} \right)_{i=1}^q ,$$

$$\text{i.e.,} \quad \boldsymbol{U} = \int_0^{2\pi} e^{-i\theta} \mathrm{d}E(\theta).$$

Then we get $(\mathbf{f}_n^{(i)}, \mathbf{f}_0^{(j)}) = \int_0^{2\pi} e^{-in\theta} \mathrm{d}(E(\theta)\mathbf{f}_0^{(i)}, \mathbf{f}_0^{(j)})$, since $\mathbf{f}_n^{(i)} = \int_0^{2\pi} e^{-in\theta} E(\theta) f_0^{(i)}$ for $i = 1, 2, \ldots, q$.

Let $F_{ij}(\theta) = 2\pi(E(\theta)\mathbf{f}_0^{(i)}, \mathbf{f}_0^{(j)})$ with $F(\theta) = ((F_{ij}(\theta)))$; we get $R(n)$ defined by $(\mathbf{f}_n, \mathbf{f}_0)$ equals $\frac{1}{2\pi} \int_0^{2\pi} e^{-in\theta} \mathrm{d}F(\theta)$ for $n \in \mathbb{Z}$. It is easy to see that $F(\theta) = 2\pi(E(\theta)\mathbf{f}_0, E(\theta)\mathbf{f}_0)$. Also F is bounded, non-decreasing and right-continuous on $[0, 2\pi]$, $F(0) = 0$ and F is uniquely determined by the function R on $\mathbb{Z}$. We call F the spectral distribution of $\{\mathbf{f}_n, n \in \mathbb{Z}\}$. It is also easy to check Cramér ([1] Thm. 5(b)) that for F satisfying the above condition, there exists a Hilbert space H and a stationary process $\{\mathbf{f}_n, n \in \mathbb{Z}\} \subseteq H^{(q)}$ such that F is the spectral distribution of $\{\mathbf{f}_n, n \in \mathbb{Z}\}$. F has a derivative, having non-negative hermitian values which belong to L^1, i.e., $\|F'\|_E$ or equivalently $\|F'\|$ (operator norm) is integrable. Here $\|A\|_E = (\mathrm{tr}(AA^*))^{1/2}$. From here on we regard F, F' as functions on unit circle which amounts to writing $e^{i\theta}$ for the argument instead of θ.

Theorem 3.1. (a) The moving average process $(\mathbf{f}_n)_{-\infty}^\infty$:

$$\mathbf{f}_n = \sum_{-\infty}^{+\infty} A_k \mathbf{g}_{n-k}, \quad (\mathbf{g}_i, \mathbf{g}_j) = \delta_{ij} G, \quad \sum_{k=-\infty}^{+\infty} |A_k \sqrt{G}|_E^2 < \infty$$

has absolutely continuous spectral distribution F such that

$$F'(e^{i\theta}) = \Phi(e^{i\theta}) \cdot \Phi^*(e^{i\theta}), \quad \Phi(e^{i\theta}) = \sum_{k=-\infty}^{+\infty} A_k \sqrt{G} e^{ik\theta} \quad a.e.$$

(b) If for this process $A_k = 0$ for $k < 0$, then

$$\Phi(e^{i\theta}) = \sum_{k=0}^\infty A_k \sqrt{G} e^{ik\theta}$$

and either $\Delta\Phi_+$ vanishes identically or $\log\Delta F' \in L^1$ on $\mathbb{T}$, the circle group, and

$$\log\Delta(A_0GA_0^*) \le \frac{1}{2\pi}\int_0^{2\pi} \log\left(\Delta(F'(\theta))\right)\mathrm{d}\theta.$$

Here $F_+(z) = \sum_{n=0}^{\infty} A_n z^n$ for $|z| < 1$ with $F_+(e^{i\theta})$ as boundary value, and Δ denotes the determinant.

Proof. (a) follows by computing $(\mathbf{f}_n, \mathbf{f}_0)$ and representing it in spectral representation. One then observes that $\sum_{k=-\infty}^{+\infty} |A_k\sqrt{G}|_E^2 < \infty$ implies that Φ, $\Phi^* \in L^2$, and the uniqueness of the Fourier transform for $\Phi\Phi^*$ gives the result. Now Φ being one-sided expandable gives that each of its elements are in the Hardy space H^2 on the circle. So $\Delta\Phi_+$ vanishes identically or $\log\Delta F' \in L^1$ and

$$\log|\Delta(A_0 G^{1/2})| = \log|\Delta\Phi_+(0)| \le \frac{1}{2\pi}\int_0^{2\pi} \log|\Delta\Phi(e^{i\theta})|\mathrm{d}\theta$$

using Szegö's Theorem. Observe $\Delta(A_0 G A_0^*) = |\Delta(A_0\sqrt{G})|^2$ and $\Delta F' = |\Delta\Phi|^2$ to complete the proof of (b).

Using the Wold decomposition and the expression for $(\mathbf{f}_n, \mathbf{f}_0) = (\mathbf{u}_n + \boldsymbol{\nu}_n, \mathbf{u}_0 + \boldsymbol{\nu}_0)$, we get that the spectral measure $F = F_u + F_\nu$ where $\{\mathbf{u}_n\}$ is purely non-deterministic and $\boldsymbol{\nu}$ is deterministic. Thus we get $F_u'(e^{i\theta}) = \Phi(e^{i\theta})\Phi^*(e^{i\theta})$, where Φ is defined from the moving average representation. F_ν is singular. If $\{\mathbf{f}_n\}$ has full rank, then by definition G is invertible, since $A_0 G = G$ as $A_0 G = (\mathbf{u}_0, \mathbf{g}_0) = (\mathbf{f}_0, \mathbf{g}_0)$ giving $A_0 = I$. Thus

$$\log\Delta(G) \le \frac{1}{2\pi}\int_0^{2\pi} \log\Delta(F'(e^{i\theta}))\mathrm{d}\theta.$$

Now we study the expression for prediction error as in the univariate case. Let

$$P(z) = \sum_{n=0}^{N} A_n z^n, \quad P(f) = \sum_{n=0}^{N} A_n f_{-n}.$$

Then $(P(f), P(f)) = \frac{1}{2\pi}\int_0^{2\pi} P(e^{i\theta})\mathrm{d}F(e^{i\theta})P^*(e^{i\theta})$ using the isomorphism between $L^2(F)$ and $\mathcal{M}_\infty$ ($f_n \to e^{-in\cdot}I$)

$$\log\Delta(P(f), P(f)) \ge \frac{1}{2\pi}\int_0^{2\pi} \log\left(F'(e^{i\theta})\right)\mathrm{d}\theta + \log|\Delta A_0|^2.$$

To see this, we first observe ([19], 4.10 and 3.11(c) of Part I)

$$\Delta\left\{\frac{1}{2\pi}\int_0^{2\pi} P(e^{i\theta})\mathrm{d}F(e^{i\theta})P^*(e^{i\theta})\right\} \ge \Delta\left\{\frac{1}{2\pi}\int_0^{2\pi} P(e^{i\theta})F'(e^{i\theta})P^*(e^{i\theta})\mathrm{d}\theta\right\}.$$

181

Since values of $P(e^{i\theta})F'(e^{i\theta})P^*(e^{i\theta})$ are non-negative definite hermitian for each θ, one has by ([19], 3.12, part I)

$$\log \Delta \left\{ \frac{1}{2\pi} \int_0^{2\pi} P(e^{i\theta})F'(e^{i\theta})P^*(e^{i\theta})d\theta \right\}$$

$$\geq \frac{1}{2\pi} \int_0^{2\pi} \log |\Delta(P(e^{i\theta}|^2 + \frac{1}{2\pi} \int_0^{2\pi} \log(F'(e^{i\theta}))d\theta.$$

As $\Delta|P(z)|$ is a polynomial in z, the first term is $\geq \log|\Delta P(0)|^2 = \log|\Delta(A_0)|^2$. Hence we have $FG'F \in L^1$ on $[a, b]$ and $\int_a^b F(x)dG(x)F^*(x) - \int_a^b F(x)G'(x)F^*(x)dx$ is non-negative hermitian (Theorem 4.10 [19]).

Wiener and Masani now get an expression for prediction error in full rank case,

$$\Delta(G) = \exp \left[\frac{1}{2\pi} \int_0^{2\pi} \log (\Delta F'(\theta))d\theta \right].$$

As a consequence, they get a purely analytic result on factorization of matrix valued functions.

Theorem 3.2. Given a non-negative hermitian matrix valued function F on $\mathbb{T}$ such that $F \in L^1$ and $\log \Delta F \in L^1$, then there exists a function $\Phi \in L^2$ on $\mathbb{T}$ with Fourier coefficients vanishing for $n < 0$, such that

$$F(e^{i\theta}) = \Phi(e^{i\theta})\Phi^*(e^{i\theta}) \quad a.e.$$

and $\Phi_+(0)$ is non-negative hermitian matrix with

$$\Delta(\Phi_+(0))^2 = \exp \left(\frac{1}{2\pi} \int_0^{2\pi} \log \Delta(F(e^{i\theta}))d\theta \right).$$

We shall call Φ satisfying the above condition a generating function. This was used by Wiener-Masani to give the form of innovations in terms of the observed process under the boundedness condition:

$$\lambda I \prec F(e^{i\theta}) \prec \lambda'I, \quad 0 < \lambda \leq \lambda' < \infty \quad \forall \theta,$$

giving an analogue of Theorem K2 in Section 1 ([19], Part II, Theorem 5.5). Here, for two non-negative definite matrices A, B, we write $A \prec B$ if $B - A$ is a non-negative definite matrix.

Independently of Wiener-Masani, H. Helson and D. Lowdenslager [5] studied the prediction problem giving the right of precedence to Wiener-Masani. They used the technique of shift invariant subspaces due to Beurling. In a subsequent paper, Masani [11] gave an "elegant and unifying" treatment of the two

182

approaches ([3]) using a generalization of a theorem of Halmos on isometries. This is generalized in ([6], [16]) for random fields and a proper generalization of Beurling theorem ([10]) follows. As stated in ([5], p.181), this was a difficult problem and the exact analogue was not possible ([14]). But Masani's paper inspired the solution.

For other influences of their work, we refer the reader to *Norbert Wiener Collected Works*, III (ed. P. Masani) and papers of H. Salehi [15] and P.S. Muhly [13].

References

[1] Cramér, H., On the theory of stationary random processes, *Ann. Math.*, **41** (1940), 215-230.

[2] Doob, J. L., *Stochastic Processes*, John Wiley, New York, 1953.

[3] Doob, J. L., Review MR0140930 (25#4344), *Math. Reviews*.

[4] Gangolli, R., Some Recollections of Pesi Masani's Years in Bombay (Pesi Masani Volume), *Prediction Theory and Harmonic Analysis*, (Ed. V. Mandrekar and H. Salehi), North Holland, Amsterdam, 1983.

[5] Helson, H. and Lowdenslager, D., Prediction theory and Fourier analysis in several variables, I, *Acta Math.*, **99** (1958) 165-202, and II, *Acta Math.*, **106** (1961), 175-213.

[6] Kallianpur, G. and Mandrekar, V., Non-deterministic random fields and Wold and Halmos decompositions for commuting isometries (Pesi Masani Volume), *Prediction Theory and Harmonic Analysis* (Ed. V. Mandrekar and H. Salehi), North Holland, Amsterdam, 1983.

[7] Kolmogorov, A. N., Stationary sequences in Hilbert space (Russian), *Bull. Math. Univ. Moscow*, **2** No.6 (1951) (English translation Natasha Artin).

[8] Kolmogorov, A. N., Interpolation and extrapolation von stationären zufälligen Folgen, *Bull. Acad. Sci. (Nauk), U.R.S.S. Ser. Math.*, **5** (1941), 3-14.

[9] Mandrekar, V., Second order processes, RM-406, Department of Statistics and Probability, 1980 (unpublished).

[10] Mandrekar, V., The validity of Beurling theorems in polydiscs, *Proc. Amer. Math. Soc.*, **103** (1988), 145-148.

[11] Masani, P., Shift invariant subspaces and prediction theory, *Acta Math.*, **107** (1962), 275-290.

[12] Masani, P. and Vijayaraghavan T., An analogue of Laurent's theorem for a simply connected region, *J. Ind. Math. Soc.*, **16** (1952), 25-30.

[13] Muhly, Paul S., The function-algebraic ramification of Wiener's work on prediction theory and random analysis, *Norbert Wiener's Collected Works*, III (Ed. P. Masani), 339-370.

[14] Rudin, W., Invariant subspaces of H^2 on a torus, *J. Functional Anal.*, **61** (1985), 378-384.

[15] Salehi, H., The continuation of Wiener's work on q-variate linear prediction and its extension to infinite-dimensional spaces, *Norbert Wiener's Collected Works*, III (Ed. P. Masani), 307-338.

[16] Slocinski, M., On Wold type decomposition of a pair of commuting isometries, *Ann. Polon. Math.*, **37** (1980), 255-262.

[17] Szegö, G., Beiträge zur Theorie der Toeplitzshen Formen, *Math. Zeitschr.*, **6** (1920), 167-202.

[18] Wiener, N., The theory of prediction, *Norbert Wiener's Collected Works*, III (Ed. P. Masani), 138-163.

[19] Wiener, N. and Masani, P., The prediction theory of multivariate stochastic processes, Part I, *Acta Math.*, **98** (1957), 111-150, Part II, *Acta Math.*, **99** (1958), 93-137, Part III, *Acta Math.*, **104** (1960), 141-162.

[20] Zasuhin, V., On the theory of multidimensional stationary random processes (Russian), *Dokl. Acad. Sci.*, **33** (1941), 435-437.

Texts and Readings in Mathematics